A Classical Philosophical Approach to Artificial Intelligence

"I found in this book an impressive intersection between Aristotelian philosophy and Artificial Intelligence. The author also mentions that, in economics, questions have been raised about why we should *build multidisciplinary paths* between Aristotle and economics. However, moving toward interdisciplinary fields and crossing previously unimagined barriers is a great way to *generate truly original ideas*. Regarding the chapters, I felt particularly connected with Chapter 5: '*Do AI Really Think?*' and I found a *rigorous explanation* of deduction, induction (possible for artificial neural networks—ANNs), and abduction (not possible for ANNs). It was also very interested to reflect on Chapter 8: '*Moral Machines?*' where the author incorporates an incompatibilist point of view on these issues—whether AI has free will or not. Apparently, it does not, but it certainly has moral implications. A book not to miss."

—Mariana Olezza, *University of Buenos Aires, Argentina*

"A lucid philosophical exploration of artificial intelligence through Aristotelian thought. It examines AI as a human artefact that must be shaped by responsibility, and ethics."

—Jaime Nubiola, *University of Navarra, Spain*

Ricardo F. Crespo

A Classical Philosophical Approach to Artificial Intelligence

Merging Technology and Humanity

palgrave
macmillan

Ricardo F. Crespo
IAE Business School
Universidad Austral
Pilar, Buenos Aires, Argentina

ISBN 978-3-032-18461-0 ISBN 978-3-032-18462-7 (eBook)
https://doi.org/10.1007/978-3-032-18462-7

This Palgrave Macmillan imprint is published by the registered company Springer Nature Switzerland AG.
The registered company address is: Gewerbestrasse 11, 6330 Cham, Switzerland

Preface

Artificial Intelligence is one of the most transformative forces of our time—pervasive, powerful, and deeply influential across virtually all dimensions in life. Yet despite its prominence, its contributions and potentially threatening effects prompt diverging appraisals. In this context, a philosophical approach certainly helps to shed clarity about what AI truly is, what it can and cannot become, and how it ought to be integrated into human society.

This book, as its title indicates, specifically adopts a classical philosophical perspective. Drawing from the enduring wisdom of thinkers like Aristotle and the long tradition of metaphysical inquiry, it asks fundamental questions: What is the nature of AI? Does it really think? Can it be free? Does it have a conscience? Can it behave morally?

These queries are not merely theoretical: Their answers have pressing ethical, social, and cultural consequences. As AI increasingly mediates our decisions, our relationships, and even our self-understanding, it becomes urgent to reflect on its boundaries and its potential through the lens of philosophy.

The decision to adopt a classical philosophical approach stems from my experience that Aristotle's and classical-oriented thinking provide clear insights into contemporary problems. Classical philosophy—metaphysics, the theory of knowledge, ethics—offers strong conceptual tools to analyse the nature of AI, its traits, and its place of in human life and society.

This book, therefore, is not intended just as an academic exercise; it contains a proposal for a dialogue to build a bridge between technology and humanity. It aims to complement the work of scientists and engineers by grounding technical progress on philosophical foundations. It also speaks to corporations, philosophers, and policymakers concerned with the deep changes AI brings for individuals and society alike.

The tone of the book is balanced, yet cautiously critical. While recognizing the genuine benefits of AI, it also highlights its likely dangers, particularly when detached from an accurate notion of its nature and lacking moral judgment as well as human responsibility. It warns about the subtle erosion of agency and meaning that can unfold when machines are mistaken for minds, and when instrumental reason undermines ethical thinking. If this work contributes to a more thoughtful engagement with artificial intelligence—one that preserves human beings' dignity and advances the moral ends of human action—then it will have fulfilled its purpose.

I would like to express my appreciation for the notes provided by Jaime Nubiola, Juan José Sanguineti, Mariano Asla, Verónica Moreno, Mariana Olezza, Alfredo Marcos, and Miguel Alfonso Martínez Echevarría on the entire book; the comments offered by Daniel Liut on Chaps. 3, 4, and 5, and by Hugo Rufiner on Chap. 3, as well as the contributions made by Alejandro Vigo on the Aristotelian ideas discussed in several chapters. I had the opportunity to discuss the whole content of the book at a Seminar held by the Institute of Philosophy (Universidad Austral, Argentina) and collected useful feedback from the audience. I am also grateful for the permission to draw from my article 'Does AI Think?,' *Scientia et Fides*, 12/2, 2024, pp. 37–47, for Chap. 5, and from my article 'Moderating the Effects of "Surveillance Capitalism": an Aristotelian Perspective,' *AI and Ethics*, 5, 2025, pp. 305–312, for Chaps. 7 and 9. I had previously presented this last work at the 8th International Conference on Applied Economics, 'Economics & the Social Sciences in the Age of AI,' Volos, Greece, on September 21–23, 2023.

My gratitude also goes to Amy Invernizzi, Senior Editor of Philosophy in Palgrave Macmillan, who enthusiastically supported this project, and to Sujatha Mani, who attentively followed its production process, while Mariana Donadini carefully revised the style of the entire manuscript.

I also want to thank the academic institutions where I work, for their unwavering support in my endeavours: Universidad Austral's IAE Business School and Philosophy Institute (Pilar, Argentina) and Argentina's National Scientific and Technical Research Council (Consejo Nacional de Investigaciones Científicas y Técnicas, a.k.a. CONICET).

Lastly, I should note that my greatest debt of gratitude is with my family, to whom I dedicate this book, as it would not have been possible without their support.

Buenos Aires and Mar del Plata, Argentina
2026

Ricardo F. Crespo

Competing Interests The author has no competing interests to declare that are relevant to the content of this manuscript.

Contents

About the Author

Ricardo F. Crespo is Professor of Philosophy of Economics at IAE Business School and at the Institute of Philosophy, Universidad Austral, and Researcher at the Argentine Council of Scientific Research (CONICET). He is a graduate in economics and philosophy and earned a Ph. D. in Philosophy (Universidad Nacional de Cuyo) and another in Economics (University of Amsterdam). He has extensively published articles and chapter books on his research topics. Recent publications of him include articles in *Synthese*, the *Journal of Institutional Economics*, *Foundations of Science*, *Cambridge Journal of Economics*, and the *Journal of Applied Economics*. His last book is The Nature of the Economy—Essays on the Philosophy and Epistemology of Economics (Palgrave Macmillan 2022). His current research interests include the philosophy of AI, explanation in the social sciences, economic rationality, and ethics in economics.

CHAPTER 1

Introduction

Abstract Given the current and future relevance of Artificial Intelligence (AI), its advantages and challenges, this book argues that it is necessary to take a philosophical look at it and its development. To that end, it adopts a classical philosophical approach, taking into account that Aristotle's views and classical thinking—metaphysics, the theory of knowledge, ethics—provide clear insights into contemporary problems. This approach includes strong conceptual tools to analyse the nature of AI, its traits, possibilities, and limitations, as well as its place in human life and society. This Introduction briefly describes the contents and conclusions of every chapter.

Keywords A classical philosophical view of AI • Current technology • AI's intelligence • AI and ethics

It is an evident, indisputable fact that Artificial Intelligence (AI) is becoming increasingly relevant in today's world. AI is everywhere, ubiquitously showing up in the most domestic affairs and the most universal events. However, 'AI' is a very generic label that encompasses a multitude of realities and fuels very dissimilar reactions. Some people share a very positive view on it, stressing the important services provided by its applications to everyday life and in a plethora of specific fields. Others, even recognizing

R. F. Crespo, *A Classical Philosophical Approach to Artificial Intelligence*, https://doi.org/10.1007/978-3-032-18462-7_1

its beneficial contributions, grimly focus on possible downsides of AI utilization and its damaging consequences. Finally, there are apocalyptic visions of the terrible future impact of AI developments (Strong and Artificial General Intelligence, AGI) on humankind.

These different takes on AI require some deep reflection to shed some philosophical light on it. A philosophical approach is needed to look at the nature of technology, including the essential characteristics of AI and its potential abilities to think and to act freely. This analysis will reveal some of the current and upcoming perils of AI, calling for an appraisal of its actual pros and cons to determine the position and actions that should be adopted in its regard. Philosophical notions drawn from metaphysics, epistemology, philosophical anthropology, and ethics will come in handy to gain a better understanding of this technological breakthrough.

A firm grasp of the nature of AI will help to make sense of the multiple current views on it. Discovering the nature of something is a metaphysical task, and, in this instance, it proves urgent, as the meaning of AI and its fundamental components remain unclear, and we need to understand them properly to deal with AI itself. This metaphysical or ontological undertaking is the first step in a probe about potential AI consequences. What a thing is, as Aristotle notes, is the starting point of any investigation (cf. Aristotle's *Analytics* I, 1, 402b16–403a2). Time relentlessly marches on; AI moves forward day by day, and we seem to run behind, forever trying to catch up.

Metaphysics is useful for all activities and sciences because it zeroes in on discerning the nature of their subject matters. As Aristotle also states in *Metaphysics*, 'For those who wish to get clear of difficulties it is advantageous to discuss the difficulties well; for the subsequent free play of thought implies the solution of the previous difficulties' (III, 1, 995a26–29). Philosophical definitions help to prevent mistakes as well as to save time by pointing directly to the kind of research that is worth doing. A steady interaction with philosophy is indispensable to every knowledge and science. As German philosopher Martin Heidegger has stated (2010, p. 214):

> Every science rests on foundations, which, considering the inquiries and methods of its manner of research, remain in principle inaccessible to the science itself. Nevertheless, every researcher is able to address these foundations thoughtfully assuming that he dares, with an alert mind, to engage in a dialogue with philosophy.

We need this conversation between philosophy and AI. The title of this book names the specific philosophical perspective that I embrace in it: classical philosophy, in reference to the school of thought of ancient Greek philosophers. As Alfred N. Whitehead argues, 'Greece was the mother of Europe, and it is to Greece that we must look in order to find the origin of our modern ideas' (1967, p. 7). He also asserts that the Greeks had an 'almost infallible instinct for hitting upon things worth thinking about' (1911, p. 128). Whitehead was not referring to contemporary Greece but to the Greek philosophical thinking that flourished over four centuries B.C., especially pointing to Aristotle. In his book *Greek Thinkers*, Austrian philosopher Theodor Gomperz also notes that 'Almost the whole of our intellectual culture is of Greek origin (...) Even those who have no acquaintance with the doctrines of the great masters of antiquity, and who have not even heard the names of Plato and Aristotle, are, nevertheless, under the spell of their authority' (1955: 528). With regard to Greek philosophers, Charles de Koninck (1964, p. 3) asserts that 'in pondering the simplest things and searching in them for the basis of whatever needs to be explained, they showed themselves to be possessors of true wisdom.' At a specific time and place in history, something extraordinary happened: men started to put myths aside and to look for rational explanations for all the things and events surrounding and inside them. This was a turning point in the evolution of mankind, and the deftest and most prominent character in this process was Aristotle.

When I mention Aristotle in discussions on economics—and now referring to modern technique and AI—people often ask, 'But what does Aristotle have to do with the economy? Wasn't technique in a clearly primitive state in his time?' Typically, I answer in an indirect fashion. Today we live in a rarified and murky intellectual atmosphere: there are too many currents and stances, partly overlapping and partly contradictory. In unclear situations like the current environment, it is helpful to search for simpler and 'less contaminated' explanations. My experience is that we find them in ancient philosophers, especially in Aristotle. Yet, that is not only my experience. Aristotelian scholar Richard Sorabji starts his classic book *Necessity, Cause, and Blame. Perspectives on Aristotle's Theory* (1980, p. ix) by stating, 'I believe and shall argue, we can benefit from going back to the views of another period, views which are sometimes refreshingly different from our own.' Explicitly, Aristotle is making a comeback in all philosophical fields— Metaphysics, Logic, Epistemology, Philosophy of Science, Political and Social Philosophy, Ethics, Philosophical

Anthropology.[1] A recent book by John Sellars on Aristotle, *The World's Greatest Philosopher*, highlights his influence on today's thinking and life: 'Aristotle's ideas and concepts have seeped into our natural ways of thinking, to the point of becoming imperceptible' (2023, p. 3).

Philosopher of technology Shannon Vallor, specifically discussing AI, states (2022, my translation):

> Aristotle and the classical philosophers talk about practical wisdom, which consists of adapting knowledge to a changing world. Our world is changing at a speed Aristotle could have never imagined, but we still need to keep this valuable lesson in mind. Because, actually, the world, human beings, and technology have always been changing. And classical philosophers have already reflected on how to adapt to an ever-changing world while preserving our virtues.

As a result of my education at a Department of Economics with philosophical leanings and a strongly Aristotelian Department of Philosophy, Aristotle has permeated my philosophical analyses. I have always found Aristotle's thinking to be especially inspiring when it comes to addressing contemporary problems. He proposed a sound philosophical system including a number of disciplines like Metaphysics, Logic, Philosophy of Science, Ethics and Politics that prove useful to approach any present reality. Then, my interest in drawing from Aristotelian notions to shed light on the nature of AI is not a matter of erudite concern but a search for what I consider smart insights and contributions to understanding AI, regardless of the novelty of this development. Almost all chapters include Aristotle's thinking on their specific topics. I try to do it in a thorough fashion, which may prove somewhat tedious for some readers. However, this rigour is intended for the Aristotelian scholars in the audience, while readers who are not interested in these in-depth analyses can skip to the final conclusive paragraph of every section.

The book will seem slightly more pessimistic than optimistic. It will acknowledge all the advantages of contemporary and future AI developments for the good of humankind and society, but it will also look at its undesirable effects. The good intentions of AI company owners, developers, and users in the so-called AI ecosystem, as well as the

[1] For the present relevance of Aristotle's thinking in all the branches of philosophy, see Stephen Brock (ed.) 2000.

declarations, regulations, and codes of ethics concerning AI, are all threatened by what Aristotle called *akrasia*—lacking willpower—or more directly by other economic and political interests that also drive AI development.[2] The strong political and economic pressures on AI intensely jeopardise any good intentions and regulations. In addition, the impact of digital society on human beings' anthropological condition, on society, and on culture has many negative consequences. However, there is hope. As I will elaborate especially in Chap. 9 and in the Conclusion, a radical change of our moral attitudes would transform the AI system in the collaborative instrument that we all desire.

The way I see it, AI will evolve technologically to such an extent that it will provide an almost perfect emulation of human thinking, actions, and emotions, but it will not be aware of all this.[3] Consciousness is a human attribute that AI will not be able to have because it is an artefact. We are aware that we are thinking, feeling, suffering. Awareness and abstraction entail a mental condition that AI lacks. Chap. 5 will explain the scope and limits of AI's 'cognitive' capabilities: it will imitate humans almost perfectly, as it will increasingly recognize human patterns and repeat them. Yet, AI will 'mirror' those patterns, as Vallor argues in her latest book (2024). A mirror does not fully reflect the thing reflected, and it can also distort it. The distinction between artificial and human intelligence will always be present, and, although in the future this distinction may prove nearly imperceptible, it will surely make a difference. As Julian Nida-Rümelin and Nathalie Weidenfeld state, 'digital states and processes *simulate* mental ones but are not identical to them, even if that simulation were perfect' (2022, p. 17). Even now, some people build emotional connections with AI robots and—mistakenly but satisfactorily—almost completely unconsciously assign emotional reciprocations to them. As Sherry Turkle has warned, 'we are easily seduced; we easily forget what robots

[2] Anu Bradford in his book *Digital Empires* (2023) extensively analyses the American market-driven, the Chinese state-driven and the European (EU) rights-driven regulatory approaches, and the conflicts between them. She favours (and I enthusiastically agree) the EU regulatory model because it is based on fundamental rights, democracy, and fairness (see Chap. 7 of this book).

[3] Concerning emulation, the Merriam-Webster Dictionary states, '*emulate* can be used as a synonym of *imitate*,' as in 'a painter who emulates her teacher's style.' But more often, *emulating* 'is about trying to equal or surpass someone you admire by striving to master what they've accomplished' (https://www.merriam-webster.com/dictionary/emulate).

are; we easily forget what we have made' (2011, p. 74). Can we feel content with this?

Except in Chap. 7, when I discuss ethical and legal regulations concerning AI, I will not refer to specific systems or institutions, as changes unfold at breakneck speed, and this book would rapidly become outdated if I did. This is one advantage of the philosophical approach: the metaphysical, epistemological, anthropological, and ethical considerations and conclusions will not change with the development of AI, once its nature is established.

Let me offer a brief overview of the content of every chapter so that readers know what to expect.

Chapter 2 deals with technology—AI is a form of it and must be construed in the context of modern technological evolution. Making a clear distinction between 'technique' and 'technology,' this chapter explains the former with the classical notion of *téchne*, a fruit of *poiesis*, a dimension of human action conceived as subordinated to *praxis*, the moral aspect of human action. In turn, because the modern concept of technology has been developed apart from morality and with a dominating attitude towards nature, the chapter stresses the need to create a 'technological humanism' (Marta Bertolaso & Alfredo Marcos, 2023) and to uphold the technomoral virtues proposed by Vallor (2016).

Chapter 3 provides a succinct account of the definition and history of AI, underscoring the idea that AI is part of a system, including human beings, institutions, and material stuff. This glance at the evolution of AI over time spans Norbert Wiener's cybernetics to today's latest developments.

Chapter 4 first deals with the characteristics of the AI system, delving into what a system is and its features. Next, it compares the man-machine and the man-AI systems, drawing from Aristotle's ideas to define the ontological nature of these systems as ordered wholes guided by common final causes. Then, it investigates the nature of AI, establishing its artefact status, as indicated by the 'A' in its name, and relying on Aristotle to explore the nature of artefacts, as accidental wholes consisting of natural elements. Artefacts differ from natural beings because they do not have an internal principle of dynamism, and their end is provided by their designers and users. Finally, this chapter characterizes AI as an abstract, yet autonomous artefact.

Noting that AI's 'intelligence' is completely different from human intelligence, Chap. 5 introduces the Aristotelian notion of intelligence, which deviates entirely from the AI concept itself. As Evandro Agazzi points out, we fall in 'the pretension of claiming that two beings are *essentially* of the same nature in they are both able to *do* a certain number of identical or very similar things (underestimating the question about *how* these two beings do the same things, and also that of the existence of *other things* which one of them may be able to do while the other is not)' (1981, pp. 196–197). Abstraction and abduction are operations that are exclusively human, also in the sense that human intelligence grows within the wholeness of human beings, including sensory, emotional, moral, spiritual, social, and contextual elements. AI robots do not have human bodies and thus they cannot experience actual sensations, pain, or physical pleasure. As posited by Aristotle, human beings can access the essences and causes of things and events, while AI only groups things under a common name and statistically infers their outcomes. In short, both activities may be regarded as 'intelligence' but only in an equivocal way.

Chapter 6 appraises the impact of AI's development on people, society, culture, economy, and politics, and the bottom line turns out to be negative. This chapter does not consider the future developments—like AGI—because these are only hypothetical today. Nonetheless, the possibility of having AI friends or AI partners—which will be neither actual friends or partners because this is something that only human beings can be—will bear unpredictable anthropological consequences.

Chapters 7, 8, and 9 address the potential remedies for AI's likely negative consequences. These chapters—especially Chap. 9—try to merge technology (AI) and humanity, as cited in the subtitle of the book.

Chapter 7 surveys the definitions, legal regulations, codes of ethics, and accountability means already established for AI. It starts considering Aristotle's ideas about the role of the State in reference to people's wellbeing, highlighting his claim that education and law promote citizens' virtues and happiness. The rest of the chapter outlines the regulations that have been developed to serve as tools to create an ethical atmosphere in the AI system. Yet, some hurdles to accountability still plague this system, given its autonomy, opacity, and the multiple actors involved in its development.

Chapter 8 analyses the possibility of developing a 'moral machine' and comes to the conclusion that, on account of its own nature, AI does not have the freedom and moral aptitude required for morality. Still, Chap. 9 stresses the need for all the people involved in the design, production, commercialization, and use of AI to have the necessary virtues to build an

AI that fosters human and social goodness. This chapter argues that a radical change in human behaviour and in the behaviour of political and economic institutions is needed to achieve collaborative action in AI's ecosystem. This behavioural change should hinge on enforcing the principles proclaimed by declarations, laws, and codes of ethics to accomplish a real ethical conduct based on ethical virtues. The above-mentioned technological humanism proposal, technomoral virtues, the digital humanism initiative (https://caiml.org/dighum/), and other individual propositions all underline the need for a thorough ethical mindset change to successfully achieve an AI development in favour of the human and social greater good. Meanwhile, as users, we should try to interact with AI exercising wisdom and prudence to maximize its benefits in the best possible way.

A short conclusion brings the book to an end, pondering AI's benefits, its threats, and the means to overcome the latter.

References

Agazzi, E. (1981). Intentionality and Artificial Intelligence. *Epistemologia, IV*(Special Issue), 195–228.

Bertolaso, M., & Marcos, A. (2023). *Umanesimo Tecnologico. Una riflessione filosofica sull'inteligenza artificiale*. Carocci.

Bradford, A. (2023). *Digital Empires. The Global Battle to Regulate Technology*. Oxford University Press.

Brock, S. (Ed.). (2000). *L'attualità di Aristotele*. Armando Editore.

De Koninck, C. (1964). *The Hollow Universe*. Les Presses de L'Université Laval.

Gomperz, T. (1955). *Greek Thinkers*. John Murray.

Heidegger, M. (2010). On the Question Concerning the Determination of the Matter for Thinking. *Epoché, 14*(2), 213–223.

Nida-Rümelin, J., & Weidenfeld, N. (2022). *Digital Humanism. For a Humane Transformation of Democracy, Economy and Culture in the Digital Age*. Springer.

Sellars, J. (2023). *Aristotle. Understanding the World's Greatest Philosopher*. Pelikan Books (Penguin Random House).

Sorabji, R. (1980). *Necessity, Cause, and Blame. Perspectives on Aristotle's Theory*. Cornell University Press.

Turkle, S. (2011). Authenticity in the Age of Digital Companions. In M. Anderson & S. L. Anderson (Eds.), *Machine Ethics* (pp. 62–76). Cambridge University Press.

Vallor, S. (2022, October 15). Interview in *La Nación*, Tenemos que evitar que la tecnología nos haga trabajar como máquinas, LA NACION. Retrieved March

12, 2025, from https://www.lanacion.com.ar/ideas/shannon-vallor-tenemos-que-evitar-que-la-tecnologia-nos-haga-trabajar-como-maquinas-nid15102022/.

Vallor, S. (2016). *Technology and the Virtues. A Philosophical Guide to a Future Worth Wanting*. Oxford University Press.

Vallor, S. (2024). *The AI Mirror. How to Reclaim our Humanity in an Age of Machine Thinking*. Oxford University Press.

Whitehead, A. N. (1911). *An Introduction to Mathematics*. William and Norgate.

Whitehead, A. N. (1967). *Science and the Modern World. Lowell Lectures 1925*. The Free Press.

CHAPTER 2

Technics and Technology

Abstract This first chapter offers a brief historical overview of the relationship between human beings and their technical outputs. It tracks the journey of the notion of technique from Aristotle's *téchne* concept to the present technology, stopping briefly at some recent milestones marked by the so-called humanities philosophy of technology and its view of technique, as well as the 'engineering philosophy of technology' corpus. While the literature on the broader philosophy of technology topic is vast, this chapter highlights the conditions required for a positive outlook on technology (Vallor, *Technology and the Virtues. A Philosophical Guide to a Future Worth Wanting*. Oxford University Press, 2016), which proves necessary to avoid both uncritical and demonizing views of technology. At the same time, it acknowledges that we are confronted with the well-known technological ambivalence—that is, the fact that technology can be used for good or for evil.

Keywords Technique • Technology

I am assuming that AI is one of the most developed technology products—at least today. Thus, the nature, possibilities, and consequences of technology deserve a first chapter in this book. The topic of technique has

R. F. Crespo, *A Classical Philosophical Approach to Artificial Intelligence*, https://doi.org/10.1007/978-3-032-18462-7_2

been tackled since the very beginnings of Western thinking. Plato made some remarks about it, and Aristotle developed a take on it throughout his writings. However, the Philosophy of Technology field largely emerged over the past century and has hugely grown since then. Nonetheless, I shall write this chapter with brevity in mind, so as to move swiftly on to the central issue of this book, the philosophy of AI.

Philosophy of Technique started growing as a systematic study in the eighteenth century. Carl Mitcham (1994) makes a distinction between 'humanities philosophy of technology' and 'engineering philosophy of technology.' Humanities philosophy of technology studies the philosophical meaning of technology and its impact on society and culture. Mitcham considers Lewis Mumford, José Ortega y Gasset, Martin Heidegger, and Jacques Ellul as paradigmatic scholars of this discipline. We may add Hannah Arendt, Hans Jonas, Günther Anders, and Jean Ladrière to this list, among others. Most of these thinkers' views on technology are predominantly negative. Instead, the 'engineering philosophy of technology' focuses on technology itself, its processes, and the nature of its products—artefacts—with a rather consistently positive outlook. This field surfaced in the past century as well, and Mitcham cites Ernest Kapp, Peter Engelmeir, Friedrich Desauer, and a few more as its leading proponents. For the sake of briefness, I will not delve into these scholars' thinking and other associated discussions,[1] concentrating instead on the similarities and differences found in the classical view of technique and the contemporary notion of technology.

Something to clarify from the start is the use of the terms 'technique' or 'technics' (I will use them interchangeably) and 'technology.' The prevalent term in today's literature on the topic, especially in English, is technology. Technique comes from the Greek term *téchne*, and it is translated into the German *Technik*, the French *technique*, the Italian *tecnica*, and the Spanish *técnica*.[2] However, these words are usually translated as *technology* in English. For instance, the English translation of Jürgen Habermas' essay 'Technik und Wissenschaft als Ideologie' (1968) is 'Technology and Science as Ideology' (1971), and the English translation of Friedrich

[1] For a summary of the topics related to Philosophy of Technology, see Franssen et al. (2024).

[2] For example, Oswald Spengler, Friedrich Jünger, Martin Heidegger, Jürgen Habermas, and Heinrich Beck use the word *Technik* for the title of their books on technique, while Jacques Ellul sticks to the French word *technique*, and José Ortega y Gasset to the term *Técnica* in Spanish.

Rapp's book *Analitische Technikphilosophie* (1978) is *Analytical Philosophy of Technology* (1981). Heidegger's 'Die Frage nach der Technik' (lecture delivered on November 18, 1955) has been translated as 'The Question Concerning Technology' (1977).

This chapter will follow Mitcham's (1994, pp. 117ff.) analysis of these terms, *téchne* (technique) and technology, to conclude that *technique* corresponds to the classical Greek and Medieval conception of technics, while *technology* leans towards a more modern take on it. Though as noted, Plato has dealt with the notion of *téchne*, I will only analyse Aristotle's thinking about it because his approach is more systematic on most topics and specifically on technique.

Aristotle's Notion of *Téchne*

Aristotle describes *téchne* as a '*héxis tis meta lógou alethous poietiké estin*'—'a state involving true reason concerned with production' (*Nicomachean Ethics*—*NE*—VI, 4, 1140a20–21).[3] *Héxis* is a habit, while *poietiké* is a kind of human action that has a transcendent result, a production. *Lógou alethous* refers to the true capture of the form or *eidos*, in which the desired outcome resides.[4]

Gianni Vattimo (1961, pp. 64ff.) has shown that art (*ars* is the Latin translation of *téchne*) or technique—*téchne*—carries two meanings for Aristotle. The most common is habit, but Aristotle also uses the term *téchne* to mean *dynamis*—the capacity or general principle of human actions—in the *Physics* and other writings. In fact, W. D. Ross translates *téchne* as 'state of capacity to make' (*NE* VI, 4, 1140a 4). Additionally, this term has other meanings that will be explored later in this chapter.

Habit (*héxis*) is a quality, a 'having' (*Metaphysics*, V, 20, 1022b 5) that proves to be a more lasting and stable quality than a mere disposition (*Categories* VIII 8b28, 9a6). It is an acquired behaviour pattern regularly followed until it becomes almost involuntary—a dominant or regular disposition or tendency.

[3] According to Terence Irwin's translation. I should make it clear that I will refer to useful *téchne*, not the mimetic *téchne* or today's art, which Aristotle deals with in his *Poetics*.

[4] *Téchne* is rational because it comes from the universal produced by experience (*empeiria*) (*Posterior Analytics* II, 19, 100a6–9). And 'art arises when from many notions gained by experience one universal judgment about a class of objects is produced' (*Metaphysics* I, 1, 981a5–7).

Capacity (*dýnamis*) refers to having an ability ('a source of movement or change,' *Metaphysics* V, 12, 1019a 15) and is also a quality. Aristotle views capacities as natural (*physikes*) (*Categories* VIII 9a 14ff.). A capacity is an ability, a potentiality, a power, or talent possessed—in this case, by a human being. Humankind is equipped with some capacities that require development, as they may be dormant or active. Other capacities are not innate and must be acquired through practice, as in the case of *téchne* (*Metaphysics* IX, 8, 1049a30–32).

Thus, it may be initially concluded that *téchne* is a habit, a lasting disposition, and a capacity to make something correctly—that is, achieving the intended result. The correct verb is to make or produce, not to do or act (in the narrow sense of acting that excludes making).

In the same *NE* chapter 4 about *téchne*, Aristotle discusses two ways of approaching contingent beings, '*kaì poietikòn kaì praktón*'—'In the variable are included both things made and things done', or 'What admits of being otherwise includes what is produced and what is achieved in action' (*NE* VI, 4, 1139b38 - 1140a1, cf. also VI, 1, 1139a5–9). He continues, 'making and acting are different'; 'production and action are different' (1140a1),[5] adding that, therefore, the rational disposition concerned with action is different from the rational disposition addressing production. The former is prudence (*phrónesis*), and the latter is *téchne. Poiesis* (production) is the activity of *téchne,* while *prâxis* is the activity of *phronesis.* Apart from external consequences, *prâxis* is characterized by an immanent impact on the very agent that acts. In turn, *poiesis* has a transitive result and a practical immanent dimension. For example, work produces an extrinsic result, but it also has an intrinsic, practical or immanent dimension—the improvement of the agent's corresponding skill.

At the same time, Aristotle made a distinction between practical and poietical (or technical) reason as guiding these kinds of human actions (in the broad sense of the term that comprises practical and poietical actions) (*Metaphysics* XI, 7, 1063b36 – 1064a15; VI, 1 1025b20–26). Poietical or technical reason drives actions towards results. Practical reason wonders how one should act to find one's own fulfilment—an ethical query—while technical or poietical reason asks what means and actions are required to achieve the desired material result.

Aristotle (*Metaphysics*, IX, 8, 1050a 23–35) believed that, although some actions are purely immanent, like seeing and theorizing, there are no

[5] Ross' and Irwin's translations respectively.

purely 'poietical' actions, because all human actions leave a footprint on people's inner self, and they are guided by ends. In other words, as mentioned, 'poietical' actions feature immanent and transient aspects. Let us hear from Aristotle:

> Intellect itself, however, moves nothing, but only the intellect which aims at an end and is practical; for this rules the productive (*poitikês*) intellect as well, since everyone who makes makes for an end, and that which is made is not an end in the unqualified sense (but only an end in a particular relation, and the end of a particular operation)—only that which is done is that; for good action is an end, and desire aims at this. (*NE* VI, 2, 1139a 35–1139b3)

In other words, the productive action is ruled by the external result sought and by the end that is the same action, an immanent aspect of it. The presence of the end—later called final cause—is present in all human actions, both practical and poietical. This final cause is ultimately human beings' good life (*eu zen*). Hence, according to Aristotle, poietical or technical activities are subordinated to human beings' moral end. Summing up, *téchne* is a capacity and habit that generates poietical (technical) actions that ought to be subordinated to the ends of practical reason, the human good.

As already noted, *téchne* is also a kind of knowledge, which comes across with striking clarity in *Metaphysics*. I have quoted (the fourth footnote in this chapter) a corresponding passage: *téchne* stems from a universal judgment derived from experience (*empeiría*). While experience zeroes in on the particular, *téchne* deals with the universal. Aristotle then compares an expert with a person possessing *téchne* (I, 1, 981a13–981b9):

> With a view to action experience [*empeiría*] seems in no respect inferior to art [*téchne*], and men of experience succeed even better than those who have theory without experience. (The reason is that experience is knowledge of individuals, art of universals, and actions and productions are all concerned with the individual; for the physician does not cure man, except in an incidental way, but Callias or Socrates or some other called by some such individual name, who happens to be a man. If, then, a man has the theory without the experience, and recognizes the universal but does not know the individual included in this, he will often fail to cure; for it is the individual that is to be cured.) But yet we think that knowledge and understanding belong to art rather than to experience, and we suppose artists to be wiser than men of experience (which implies that Wisdom depends in all cases

> rather on knowledge); and this is because the former know the cause [*aitían*], but the latter do not. For men of experience know the thing is so, but do not know why, while the others know the 'why' and the cause. Hence we think also that the masterworkers [*architéktonas*] in each craft are more honourable and know in a truer sense and are wiser than the manual workers, because they know the causes of the things that are done (…); thus we view them as being wiser not in virtue of being able to act, but of having the theory for themselves and knowing the causes. And in general it is a sign of the man who knows and of the man who does not know, that the former can teach, and therefore we think art more truly knowledge [*epistémen*, science] than experience is; for artists can teach, and men of mere experience cannot.

Thus, *téchne* is also science in the sense of being a 'demonstrative habit' (*NE* VI, 3, 1139b32)—a knowledge of causes—but not a strict Aristotelian science—i.e., a science not of the necessary, but of the contingent. This does not prevent technicians from also having the experience of their *téchne*, which turns it into a capacity.

What causes underlie *téchne*? *Téchne* is a kind of knowledge and comes from knowledge, 'from art proceed the things of which the form (*eîdos*) is in the soul of the artist' (*Metaphysics* VII, 7, 1032b1–2). Thinking (*noéseos*) stems from form, and production (*poíesis*) (id. 16–17) derives from thinking, but choice is always necessary. Aristotle states: 'The origin of action [*práxeos*]—its efficient [*kínesis*], not its final cause [*héneka*] —is choice [*proaíresis*], and that of choice is desire [*hórexis*] and reasoning [*lógos*] with view to an end' (*NE* VI, 2, 1039a30–31). This latter end is the purpose of an action, and it is also the final end of *poiesis*. As already quoted, the intellect moves towards an end—the specific end of production—but it ultimately moves towards the very action in its immanent sense, which is the final end of an action (*NE* VI, 2, 1139a 35 - 1139b3). The transitive aspect of *poiesis* that produces a result is complemented by the immanent character of *téchne*, its *practical* dimension—a *héxis* that becomes more perfect thanks to a right *poiesis*. In addition to referring to the perfection of the result, the perfection of *téchne* primarily hinges on the right exercise of its *poiesis*. That is, here is a double teleology in the case of the act of *téchne*, the *poiesis*: the extrinsic end (that which is produced, its transitivity or *kínesis*) and the greater perfection of the *héxis*—i.e., of the *téchne*.

The extrinsic end is the end to which Aristotle refers when affirming that there is 'an excellence in art' (*NE* VI, 5, 1140b21), because we can check if the result is the desired result. It must match the form or *eidos* that lies in the producer's soul (mind): as quoted earlier, *téchne* is a *héxis meta lógou alethous.* According to Mitcham (1994, p. 128), the role of *logos* is to truly capture the form, which must be received by a corresponding adequate matter to generate the result. Aristotle notes, 'Since anything which is produced by something (and this I call the starting-point of the production), and from something (and let this be taken to be not the privation but the matter)' (*Metaphysics* VII, 8, 1033a23–26). Then, he explains that he means 'the matter which governs the production in the making and producing of any work of art' (*Metaphysics* VII, 9, 1034a10–11)—that is, it must be an adequate matter. In his *Physics* (II, 9, 200a11–12), Aristotle cites the example of a saw that must be made of iron.

Summing up, *téchne* is a capacity and habit that generates technical actions oriented to the production of a desired outcome that ought to be subordinated to the ends of practical reason, the human good.

From *Téchne* to Technology

When discussing the role of *logos* in *téchne*, Mitcham points out the difference between *téchne* and technology. *Logos* in *téchne* grasps the form, he asserts, but it does not constitute the *logos* of the producing activity. Mitcham asks (1994, p. 128):

> But is this not precisely what modern technology proposes to furnish—a *logos* of the activity, a rationalization of the process of production, independent of, if not actually divorced from, any particular conception of *eidos* or form?

Moreover, he wonders if this is not the source of technology's aspiration for neutrality. While *téchne* has a practical, immanent dimension, the technological process regards the role of reason as a rationalization of the process of production.

At the root of the difference between *téchne* and technology, Mitcham argues, there is an ontological conception of matter. For Aristotle, matter is not purely neutral; it is intrinsically linked to a specific form, as mentioned above. Mitcham points out that the Western ontology of matter radically changed in the seventeenth century, when it was construed as an inert thing, pure extension, without an orientation towards anything, 'something to do with as one pleases' (1994, p. 133).

Another difference concerns the relation between technique and nature. For Aristotle, '*téchne* imitates nature' (*Physics* II, 2, 194a22) and 'on the basis of Nature, carry things further than Nature can' (*Physics* II, 8, 199a16). *Téchne* is a continuation of nature, making its capacities (*dynameis*) explicit. Technique supplements and completes nature to suit human purposes (Schummer, 2001), while modern technology has a *pathos* of dominance of nature. Francis Bacon, often considered the founder of empirical science, writes in his *Novum Organum* that the highest and most excellent level of power is man´s domination of nature (pp. 105–106; 1620; part I, section 129). This view has eventually led to today's environmental issues.

There are pessimistic and optimistic perspectives about the technique-nature relationship. Pessimistic outlooks emphasize technology's attempt to overpower nature, which impoverishes humanity and destroys the planet, whereas optimistic views stress the contribution of technology to human life. Friedrich Rapp (1981) has found a way to strike a balance: he calls the technical environment created by technique 'technosphere,' and he states that we must 'pay' the positive achievements of technology by losing 'an intelligible and immediately significant lifestyle' (1981, p. 122).[6] The unnatural trait of the technosphere lies in its *dominating role* (*ibid.*): nature is no longer a structured being but a mere conglomeration of lifeless matter obeying mathematical laws and arbitrarily useable (1981, pp. 102ff. and 124). All this is part of the Weberian 'disenchantment of the world' (Max Weber, 1970, p. 155; Rapp, 1981, p. 179).

Modern technology does not factor in the intrinsic end of *poiesis* and *téchne*. The two dimensions that make up technology are instrumentality and productivity (Franssen et al., 2024, p. 5), while the immanent practical dimension of *poiesis* disappears. For Aristotle, the extrinsic end of *poiesis* and *téchne* was subordinated to the end of *prâxis*, but this subordination changes in modern times. Since David Hume, practical rationality has been reduced to instrumental rationality (see Zafirovski, 2003; Boudon, 2004, p. 57). Hume sees no rational deliberation about ends and views rumination on means as unconnected to any rational consideration of ends. His well-known statement reads, 'Reason is and ought only to be the slave of passions and can never pretend to any other office than to serve and obey them' (Hume [1739–1740] 1968, p. 415 -II, iii, 3). For him, actions are motivated by ends determined by passions, not by reason

[6] Cf. also 144–145, where he speaks of a 'technology-related alienation' and 165ff.

(id, p. 415). In other words, the sole role of reason in human behaviour is instrumental—the extrinsic end replaces the intrinsic end, which disappears. While for Aristotle the rationality of *téchne* was intrinsically subordinated to practical rationality, which is ethical, for modern technology ethics becomes an external limit at best, even a new ethics. This is because what is technically feasible becomes ethically mandatory (Ladrière, 1977, pp. 91, 108–109; Jacques Ellul, 1990, pp. 106–135). Rapp complains about 'a *technological imperative* according to which all possibilities for action must be realized simply because they exist' (1981, p. 130). Erich Fromm, who regards this as the first principle of the technological society, puts it this way, 'something *ought* to be done because it is technically possible to do it' (1968, p. 33, cursive in the original).

There is also a change in the concept of theoretical science. While classical theoretical science was speculative knowledge looking for truth, contemporary science is more a system of action close to technology than a method to acquire knowledge (see Ladrière, 1977, p. 11).[7] Ladrière states (1977, p. 11):

> the actual dynamics of science has completely overturned the traditional Western conception of reason, truth, the relationship between theoretical reasoning and practical reasoning, the purpose of man and the nature of historical reality.

Today scientific knowledge is not sapiential but operative, and technology is closely associated with this pragmatical science. Science contributes to technology, providing the necessary knowledge to transform reality (Ladrière, 1977, p. 42). The technological task involves getting the desired results with the greatest efficiency (Rapp, 1981, pp. 137, 142).

Another characteristic of the science-technology duo is its tendency to increasing self-direction toward ends, autonomy, and self-organizing. Thus, it is 'more and more foreign to humanity, and becomes a sort of external force trying to impose its own law on mankind' (Ladrière, 1977, p. 53; see also 107; see also Rapp, 1981, p. 135). Without an end beyond the desired product (as the internal end of *téchne*), science and technology become end in themselves, devoid of intrinsic ethical criteria.

[7] In fact, Mario Bunge (1966) uses the terms 'technology' and 'applied science' as synonymous.

At the same time, nobody can deny the advantages of this partnership, how everyday life has become easier. Yet, as Rapp states, 'technology serves man only to the extent that man serves technology' (1981, p. 145),[8] whereas Heidegger affirms that 'conspicuous successes of the inexorable development of technicity continue to give the appearance that the human being is the master of technicity. In truth, however, he is the servant of the power that thoroughly dominates all technical production' (2010, p. 218).[9] Furthermore, this process seems unstoppable. Alas, what should we do? How can we fix this?

Looking Ahead

The science and technology race is extraordinarily fast-paced and overwhelming—neither has any other ends but their own unlimited progress. We urgently need to reflect on them, to reintroduce practical reason into the process. The final goals cannot be defined by science and technology in their current state. Ladrière highlights the role of culture as provider of roots and ultimate ends (1977, p. 146). The true purposes of action pertain to an ethical order.

Rapp takes it for granted that the development of technology is irreversible and suggests gradually changing its orientation. A starting point is to replace the *technological* values for *human* values (Rapp, 1981, p. 176). Like Ladrière, he also stresses the subordinating role of social and cultural criteria.

There are many initiatives to regulate the increasing advancement of technology in order to obey ethical criteria: codes of ethics, specific laws, organizations devoted to control the effects of technology, etc. The AI ecosystem requires systemic regulations (cf. Chap. 7), but these means will be ineffective if there is not a thorough, overarching ethical background.

Marta Bertolaso and Alfredo Marcos (2023, p. 14 and Chap. 3) posit the notion of 'technological humanism,' meaning to put technology in the context of human life to serve people and the common good.

[8] ChatGPT serves people but to the extent that it is served by them, as they supply new information when interacting with it.

[9] Heidegger stated this on October 30th, 1965, in a lecture about the replacement of philosophy by a new science, cybernetics. He said that human beings are 'a disturbing factor' in the cybernetic calculation (2010, p. 216). As already mentioned, Heidegger had a negative view of technique—see Heidegger (1977). Lewis Mumford (1970, pp. 164ff.) speaks of a 'pentagon of power': political absolutism, energy power, productivity, profit, and publicity.

Technology should foster life and health, alleviate our vulnerability, promote forms of peaceful and fair coexistence, driving us closer to the truth, to goodness, and to beauty. Before moving on to a new form of technology, we should check that it promotes all these ends (2023, p. 91). This proposal builds on the previous development of a metaphysical approach to technology and a philosophical anthropology study of human beings.

It is interesting to cite Shannon Vallor's (2016) suggestion to develop 'technomoral virtues.' For her, the current impact of technology is something completely new, which causes a deep transformation in the way people live and in the entire relationship between human beings and technology. Vallor argues that we need to cultivate personal virtues (2016, p. 10):

> The technomoral virtues, cultivated through the practices and habits of moral self-cultivation that we can learn from the classical virtue traditions examined in this book, are humanity's best chance to cope and even thrive in the midst of the great uncertainties and vicissitudes of technosocial life that lie ahead. This hope will only be realized, however, if these virtues are more consciously cultivated in our families, schools, and communities, supported and actively encouraged by our local and global institutions, and exercised not only individually but together, in acts of collective human wisdom. This is a tall order, but not beyond our capabilities.

It seems safe to conclude that, apart from other means, virtue proves indispensable, and it is fostered by education and law. I will revisit these ideas in Chap. 9.

References

Aristotle. (1925). *Nicomachean Ethics* (Sir D. Ross Trans.). Oxford University Press.

Aristotle. (1953). *Metaphysics* (Sir D. Ross, Trans.). Oxford University Press.

Aristotle. (1994). *Posterior Analytics* (J. Barnes, Trans., 2nd ed.). Oxford University Press.

Aristotle. (1999). *Nicomachean Ethics* (T. Irwin, Trans., 2nd ed.). Hackett Pub. Co.

Bacon, F. ([1620] 1911). *Novum Organum* (Ed. J. Devey). P. F. Collier and son.

Bertolasso, M., & Marcos, A. (2023). *Umanesimo Tecnologico. Una riflessione sfilosoficca sull'imteligenzia artificiale*. Carocci.

Boudon, R. (2004). Théorie du choix rationnel, théorie de la rationalité limitée ou individualisme méthodologique: que choisir ? *Journal des Economistes el des Etudes Humaines, 14*(1), 45–62.

Bunge, M. (1966). Technology as Applied Science. In F. Rapp (Ed.), *Contributions to a Philosophy of Technology* (pp. 19–39). Reidel.
Ellul, J. (1990). *La technique ou l'enjeu du siècle*. Economica.
Franssen, M., Lokhorst, G.-J., & van de Poel, I. (2024). Philosophy of Technology. In E. N. Zalta & U. Nodelman (eds.), *The Stanford Encyclopedia of Philosophy* (Fall Edition). https://plato.stanford.edu/archives/fall2024/entries/technology/.
Fromm, E. (1968). *The Revolution of Hope. Towards a Humanized Technology*. Bantan Books.
Habermas, J. (1968). *Technik und Wissenschaft als 'Ideologie'*. Suhrkamp.
Habermas, J. (1971). Technology and Science as "Ideology". In *Toward a Rational Society* (pp. 56–87). Heinemann.
Heidegger, M. (1977). The Question Concerning Technology. In *The Question Concerning Technology and Other Essays* (W. Lovitt, Trans.). Garland Pub. Inc.
Heidegger, M. (2010). On the Question Concerning the Determination of the Matter for Thinking. *Epoché, 14*(2), 213–223. Translalted by Richard Capobianco and Marie Göbel.
Hume, D. ([1739–1740] 1968). *A Treatise of Human Nature* (L. A. Selby-Bigge, Ed.). Oxford University Press (reprinted).
Ladrière, J. (1977). *The Challenge Presented to Cultures by Science and Technology*. UNESCO.
Mitcham, C. (1994). *Thinking Through Technology. The Path between Engineering and Philosophy*. The University of Chicago Press.
Mumford, L. (1970). *The Myth of the Machine. The Pentagon of Power*. Harcourt, Brace, Jovanovich.
Rapp, F. (1981). *Analytical Philosophy of Technology* (Boston Studies in the Philosophy of Science, R.S. Cohen & M. W. Wartofsky, eds. Vol. 63). D. Reidel Publishing Company.
Rapp, F. (1978). *Analytische Technikphilosophie*. Verlag Karl Alber.
Schummer, J. (2001). Aristotle on Technology and Nature. *Philosophia Naturalis, 38*, 105–120.
Vallor, S. (2016). *Technology and the Virtues. A Philosophical Guide to a Future Worth Wanting*. Oxford University Press.
Vattimo, G. (1961). *Il concetto di fare in Aristotele. Università di Torino.* Pubblicazione della Facoltà di Lettere e Filosofia.
Weber, M. ([1919] 1970). Science as Vocation. In H. H. Gerth & C. Wright Mills (Eds.), *From Max Weber. Essays in Sociology*. Routledge and Kegan.
Zafirovski, M. (2003). Human Rational Behavior and Economic Rationality. *Electronic Journal of Sociology*, 11–13. Retrieved February 8, 2024, from www.sociology.org/content/vol7.2/02_zafirovski.html.

CHAPTER 3

AI Definitions and Key Milestones

Abstract This chapter briefly introduces the meaning and interactions of some related realities: computers, AI, algorithms, big data. It underscores that AI must be construed in the context of a system including people and human institutions. There is also a short survey of AI history to paint a broad-stroke picture of the different kinds of AI developments thus far.

Keywords Algorithm • Big Data • AI • Cybernetics • AI History

First and foremost, we should realise that we cannot look at AI on its own; rather we should regard it as a whole system that inevitably requires human beings' involvement. AI does not make sense and cannot exist without the people who create it, convey it, use it, interpret it, and interact with it. Marta Bertolaso and Alfredo Marcos (2023) speak about intelligent systems made up by individuals and machines. The entire system is intelligent because—truth be told—its intelligence component is not artificial but human, and the artificial part is not intelligent. Without a person in the loop, the system lacks 'intentionality,' which is necessary for the very system to exist. I will expound on this in Chap. 5, but, for now, it is important to understand that AI is not just an algorithm that processes data and produces information or a set of results. It needs to be contained in

R. F. Crespo, *A Classical Philosophical Approach to Artificial Intelligence*, https://doi.org/10.1007/978-3-032-18462-7_3

physical devices that run on energy. As Kate Crawford puts it, 'At a fundamental level, AI is technical and social practices, institutions and infrastructures, politics and culture' (2021, p. 8).

Three core notions are intertwined: algorithms, big data, and artificial intelligence. According to Tarleton Gillespie (2014, p. 167), 'in the broadest sense, [algorithms] are encoded procedures for transforming input data into a desired output, based upon specified calculations. The procedures name both a problem and the steps by which it should be solved.' Computers use algorithms to encode those procedures, while algorithms support AI, providing its foundations.

Large computers can process a huge amount of data, commonly called 'big data,' taken from a variety of sources (social media, mobile phones, cameras, etc.), which enables them to detect patterns, find solutions, generate opportunities, identify best practices, etc. These computers have specific features: storage capacity for large data volumes (also accommodating a variety of data formats) and high processing speed. Computers are the new 'old machines,' while algorithms and AI are the new oil or electricity. Andrew Ng says that AI 'is the new electricity' (2017): just as electricity revolutionized the world 100 years ago, AI will and is doing it now.

Artificial intelligence comprises computational devices that employ algorithms to try to simulate human intelligence. Marvin Minsky describes AI as 'the science of making machines do things that would require intelligence if done by men' (1968, p. v). The definitions are multiple because there are several types of AI, and they are continuously evolving. However, the central idea is that AI seeks to mimic human thinking—or, at least, human thought processes. Besides, 'singularity' refers to a supposed break that will happen if AI ever outdoes human intelligence.

Let us take a look at the history of these new technologies.[1]

In 1948, mathematician, philosopher, and engineer Norbert Wiener published his book *Cybernetics: Or Control and Communication in the Animal and the Machine*.The subtitle provides a definition of cybernetics, describing it as having a goal and taking action to achieve that goal, which requires communication. In the Introduction to the first edition of the book, Wiener wrote ([1948] 2019, p. 18):

[1] For a more thorough chronicle of the history of AI, see Michael Wooldridge (2020). For shorter accounts, see Michael Haenlein and Andreas Kaplan (2019) and Eric Horvitz and Tom M. Mitchell (2024) articles.

> We [the author and his colleague Arturo Rosenblueth] have decided to call the entire field of control and communication theory, whether in the machine or in the animal, by the name Cybernetics, which we form from the Greek χυβερνήτηs or steersman. In choosing this term, we wish to recognize that the first significant paper on feedback mechanisms is an article on governors, which was published by Clerk Maxwell in 1868 and that governor is derived from a Latin corruption of χυβερνήτης. We also wish to refer to the fact that the steering engines of a ship are indeed one of the earliest and best-developed forms of feedback mechanisms.

The book includes a complete chapter (5) on 'Computing Machines and the Nervous System.' In the Introduction, Wiener stated, 'It has long been clear to me that the modern ultrarapid computing machine was in principle an ideal central nervous system to an apparatus for automatic control' ([1948] 2019, p. 39). In 1950, in a new book called *The Human Use of Human Beings: Cybernetics and Society*, Wiener pointed out, 'The nervous system and the automatic machine are fundamentally alike in that they are devices which make decisions on the basis of decisions they made in the past' ([1950] 1989, p. 33). He even referred to machine learning in that same book ([1950] 1989, p. 66):

> On the one hand, I wish to give a method of constructing learning machines, a method which will not only enable me to build certain special machines of this type, but will give me a general engineering technique for constructing a very large class of such machines. [...]. On the other hand, I wish to describe such machines in terms which are not too foreign to the actual observables of the nervous system, and of human and animal conduct.

However, while cybernetics deals with self-regulating and self-organizing machines, AI's subject matter zeroes in on building an intelligent machine that equals or exceeds human intelligence. When it comes to AI specifically, there are two approaches: one based on symbolism and programming (logic-based and probabilistic) and the other one revolving around the creation of artificial neural networks (ANNs) or connectionism.

Regarding the first approach, in October 1950, Alan M. Turing published an essay entitled 'Computing Machinery and Intelligence,' in which he discussed the possibility that a machine might think in the way human thinking is typically construed. When he wrote about a machine with the ability to compute, Turing was referring to the computational systems that were emerging at that time. After analysing the potential objections to achieving that goal, he came to a positive conclusion. He had previously (1936) come up a mathematical system—currently known as a universal Turing machine—that performs every possible computation by combining binary symbols.

In that piece, Turing proposed a test to determine if a computational system displayed a behaviour that would be deemed intelligent by comparing it with an example, which in this case was a human being. Essentially, the test involved two participants—a system and a person—and an interrogator was added to asks questions to both participants without knowing which was human. The interrogator's task was to study the answers to the questions asked to determine which participant was a human being. If it were not possible to tell the two interrogated subjects apart, the machine would have passed the test. Thus, the test actually assessed the machine's ability to provide answers that resembled those given by a human would give.

Later, in 1955, John McCarthy, Marvin Minsky, Nathaniel Rochester, and Claude Shannon made a bid to hold a workshop the following year at Dartmouth College, New Hampshire, USA. In their proposal, the expression 'Artificial Intelligence' was used for the first time, referring to as intelligence deployed by a computational system. Pitching this 'summer research project,' McCarthy stated that 'the artificial intelligence problem is taken to be that of making a machine behave in ways that would be called intelligent if a human were so behaving' (McCarthy et al., 1955, p. 11). Years later, McCarthy noted that it would have been better to use the term 'Computational Intelligence,' since it would have been more accurate, but the original term prevailed and is the one we use today. This workshop included the introduction of a program intended to emulate humans' problem-solving skills, called *Logic Theorist* and developed by Allen Newell, Cliff Shaw, and Herbert Simon.

Some years earlier, in 1943, Warren McCulloch and Walter Pitts laid the foundation for artificial neural networks (ANNs) by drawing parallels between the human brain and computing machines. They developed a computational model for neural networks relying on mathematics and algorithms. In a later paper (Pitts and McCulloch 1947), they argued that thermodynamics is closer than logic when it comes to the brain's activity. Also, psychologist Daniel Hebb (1949) formulated a theory of learning based on neural interaction now known as 'Hebbian learning,' thus building a bridge between psychology and physiology, effectively paving the way for today's artificial neural networks.

In 1951, Marvin Minsky created the first artificial neural network capable of artificial learning, which was called SNARC (Stochastic Neural Analog Reinforcement Calculator). Using analogue and electromechanical components, 40 neurons were manufactured and interconnected in a network. Each neuron was designed with a capacitor for short-term memory and a potentiometer for long-term memory. This mechanism gave the machine positive rewards to learn how to get out of an imaginary maze, making it possible to experiment with the notion of automatic learning—or, as we call it now, Machine Learning. In 1958, Frank Rosenblatt published 'The Perceptron: A Probabilistic Model for Information Storage and Organization in the Brain' and introduced an artificial neural network he named 'perceptron.' Unfortunately, due to Minsky and Seymour Papert's (1969) criticism of the perceptron, the funding for the neural network program was cut off until the 1980s, when the development of probabilistic methods to face uncertainty started as well, with a non-logic-based symbolic approach.

Eventually, the idea that humans and machines are similar became more popular in the research community, prompting scientists to continue working on what it would take to develop machines with human-like intelligence. Still, it was necessary to wait until computers evolved further and were able store and retrieve information fast enough. As we know, this goal has been greatly advanced, and computers' capabilities grow every year. The history of contemporary AI forms has had its ups and downs due to technical and funding factors. Currently, funding largely comes from for-profit corporations.

We are witnessing a boom in AI research, development, and production, with machine learning standing out as a common approach. This technology enables the use of massive amounts of data (or Big Data) managed by powerful computers. An artificial neural network simulates and emulates the activity of a biological neural system, which is argued to operate not in a symbolic but in a 'connectionist' or 'synaptic' way. A classical, straightforward neural network consists of a multi-layer network of artificial neurons with an input layer, hidden layers in between, and an output layer. In such a network, each neuron processes input from lower-level neurons, forwarding its output to other neurons in the upper layer. One subset of machine learning is deep learning, with more than three layers—that is, there are more than one 'hidden' layer. Deep learning can process huge amounts of data, and, when working in operational mode, data is fed to the input layer, which is forwarded and processed all the way to the output layer to yield an AI outcome. During the learning phase (or learning mode), subsets of processed 'training data' may be compared with the desired outputs, and the corresponding errors are back-propagated (with the information flowing in only one direction from the output to the input layer) recurrently (with feedback connections) in a 'convolutional' (specialised in processing images) and reinforced manner, as the machine learns from its hits and misses. The learning process may be supervised—by a programmer that trains the system, setting outcomes, entering inputs, and providing feedback—or not—the system finds patterns in data for itself.

Margaret Boden (2016, pp. 81ff.) underscores the traits of one type of artificial neural networks that does 'Parallel Distributed Processing' (PDP): these PDP-doing artificial neural networks recognize patterns and their relations by showing examples, while recognizing incomplete or partly damaged patterns and dealing with overlapping sets of family resemblances. She also cites a tendency to develop hybrid systems, including PDP and symbolic modules (2016, pp. 97ff.). Selmer Bringsjord and Naveen Sundar Govindarajulu (2018, p. 18) refer to a triad composed by logicist, probabilistic, and neurocomputational techniques.

Usually, AI types are classified into narrow or weak and general or strong. Although some authors view strong AI and general AI as synonymous, this is not so: strong AI has a philosophical background and refers to (hypothetical) AI systems that can grow a mind and mental states, which ultimately implies consciousness. Instead, general AI has a more pragmatic motivation regarding the type and variety of tasks it can undertake without having been specifically trained, which seems to be much more realistic than strong AI (if this were actually possible).

Narrow AI developments perform specific tasks: nearly all of today's applications involve narrow AI, including Siri, Google translator, and a plethora of apps all around us at all times. A special effort currently goes into developing context-adaptive, sustainable AI applications in specific fields, featuring situational awareness. In turn, general AI can learn and function in ways analogous to human beings: with similar vision, understanding, learning, and speaking capabilities. The development of this type of AI is currently under way. As Boden (2016, p. 7) explains, it will require a deep understanding of the computational architecture of our minds. She remarks that supercomputers are not enough, and new problem-solving methods are needed (2016, p. 22). Specifically, Boden mentions heuristics, which guide the program toward specific search spaces; planning or defining a sequence of actions to achieve a final goal; mathematical simplification, used once the search space is outlined; and knowledge representation, which involves presenting the problem in a way the machine can process. Most recently, large language models like ChatGPT have given rise to generative AI programs. They are so called for their abilities to not only process data, but to generate new outputs and content. AI history is clearly not over… rather, it continues to unfold very rapidly.

References

Bertolaso, M., & Marcos, A. (2023). *Umanesimo Tecnologico. Una riflessione filosofica sull'inteligenza artificiale*. Carocci.

Boden, M. (2016). *AI, its Nature and Future*, Oxford University Press.

Bringsjord, S., & Govindarajulu, N. S. (2018). Artificial Intelligence. *Stanford Encyclopedia of Philosophy*. https://plato.stanford.edu/entries/artificial-intelligence/.

Crawford, K. (2021). *Atlas of AI*. Yale University Press.

Gillespie, T. (2014). The Relevance of Algorithms. In T. Gillespie, P. J. Boczkowski, & K. A. Foot (Eds.), *Media Technologies: Essays on Communication, Materiality, and Society*. The MIT Press.

Haenlein, M., & Kaplan, A. (2019). A Brief History of Artificial Intelligence: On the Past, Present, and Future of Artificial Intelligence. *California Management Review*, *61*, 1–10.

Hebb, D. (1949). *The Organization of Behavior*. Wiley.

Horvitz, E., & Mitchell, T. M. (2024). Scientific Progress in Artificial Intelligence: History, Status, and Futures. In K. H. Jamieson, A.-M. Mazza, & W. Kearney (Eds.), *Realizing the Promise and Minimizing the Perils of AI for Science and the Scientific Community* (pp. 147–193). University of Pennsylvania Press. https://erichorvitz.com/AI_Overview_History_Status_Futures_February_2024.pdf.

McCarthy, J., Minsky, M., & Rochester, N. (1955, August 31). *A Proposal for the Dartmouth Summer Research Project on Artificial Intelligence*. Retrieved June 4, 2024, from http://jmc.stanford.edu/articles/dartmouth/dartmouth.pdf.

McCulloch, W. S., & Pitts, W. (1943, December 1). A Logical Calculus of the Ideas Immanent in Nervous Activity. *Bulletin of Mathematical Biophysics*, *5*(4), 115–133.

Minsky, M. (1968). *Semantic Information Processing*. MIT Press.

Minsky, M., Papert, S., & A. (1969). *Perceptrons: An Introduction to Computational Geometry*. MIT Press.

Ng, A. (2017, March 11). Why AI is the New Electricity. *Stanford Graduate School of Business*. https://www.gsb.stanford.edu/insights/andrew-ng-why-ai-new-electricity.

Pitts, W. H., & McCulloch, W. S. (1947). How We Know Universals: The Perception of Auditory and Visual Forms. *Bulletin of Mathematical Biophysics, 9*, 127–147.

Rosenblatt, F. (1958). The Perceptron: A Probabilistic Model for Information Storage and Organization in the Brain. *Psychological Review, 65*(6), 386–408.

Turing, A. M. (1936). On Computable Numbers with an Application to the Entscheidungsproblem. *Proceedings of the London Mathematical Society*, Series 2, 42/3 and 42/4.

Turing, A. M. (1950). Computing Machinery and Intelligence. *Mind, 59*, 433–460.

Wiener, N. ([1950] 1989). *The Human Use of Human Beings: Cybernetics and Society*. Free Association Books.

Wiener, N. ([1948] 2019). *Cybernetics: Or Control and Communication in the Animal and the Machine*. The MIT Press.

Wooldridge, M. (2020). *A Brief History of Artificial Intelligence*. Flatiron Books.

CHAPTER 4

The Ontology of AI

Abstract The aim of this chapter is to examine the ontological nature of AI based on Aristotle's metaphysical categories. Before introducing the latter and applying them specifically to artifacts, including AI, the chapter studies the nature of the whole AI system, including algorithms, as well as the social structures surrounding its development to argue that AI is an 'abstract artifact.' As the very term 'artificial' indicates, AI is a technical artifact. What is the ontological character of artifacts and specifically of AI? This is not an easy topic to address. From Aristotle's time to the present, artifacts have been distinguished from natural things, but a 'continuum problem' emerges when dealing with natural and artificial things, and a growing body of literature delves into this issue.

Keywords The AI system • Nature of artifacts • Nature of AI

In the preceding chapters, I have upheld that we must look at the whole system of AI. We cannot think of AI as isolated from the people who design, use, interpret, and interact with it.[1] The man-machine—including

[1] John R. Searle has argued that computers and AI are observer-relative entities. See, for example, 1994, Chapter 9. He claims that even a syntactic property does not pertain to a

R. F. Crespo, *A Classical Philosophical Approach to Artificial Intelligence*, https://doi.org/10.1007/978-3-032-18462-7_4

AI—relationship has to be approached systemically. In this chapter, I will first analyse the ontological nature of the AI system and then the ontology of AI itself.

The Human-AI system

Pioneered by Ludwig von Bertalanffy, the 'General System Theory' and its key underlying notions were widely adopted by contemporary scientists to analyse natural, biological, human, social, and artificial phenomena. For this theory, 'the whole is more than the sum of its parts' and has its own ontological nature. Thus, the systemic approach propelled a real revolution that left the previously dominant mechanistic view behind. Von Bertalanffy relentlessly criticised the prevailing mechanistic analysis of his time, showing its failures across all fields—physical, biological, human, and social (see, e.g., [1968] 1973, pp. xii, 8, 23). Instead, he championed a systemic 'perspectivism' to replace the physical 'reductionism' of the mechanistic paradigm ([1968] 1973, p. 48).

In a way, the systemic approach marks a return to Aristotle's holistic views of the cosmos, nature, and human society. Aristotle regarded these realities as 'wholes'— substantial or accidental, open or closed wholes, depending on whether they had any links to external phenomena or not.[2] For him, social realities were 'accidental' open wholes, as social systems interacted with 'external' elements, like other social systems, geography, or the climate. Accidental does not mean arbitrary: for instance, Aristotle believed social community was not a substantial whole. He considered it as an accidental whole consisting of people, but he did not view it as a mere aggregate of individuals. It was a whole organised towards a shared common end. Specifically referring to the nature of society and contrary to Plato's monistic notion (*Politics* II, 2), Aristotle denied that the *polis* was a substantial unity and rejected the sophist concept of the *polis* as just an accidental unity (*Politics* III, 3). Indeed, Aristotle's rich metaphysics provides categories that allow for an 'intermediate' stance.

In Aristotelian thought, groups of things or wholes may be substantial, merely accidental, or accidental unities of order. We have a substantial

physical entity per se—it is assigned to it by the observer. This issue will not be tackled here, though.

[2] The classification of open and closed systems has been drawn from von Bertalanffy ([1968] 1973, pp. 38–40), who cites examples from physics.

unity when, for example, we combine copper and iron, and, as a result, we get bronze—each former substance loses its own substantiality, and a new substance emerges. Substance is the strongest whole. Aristotle states, 'Now most things are called one because they either do or have or suffer or are related to something else that is one, but the things that are primarily called one are those whose substance is one' (*Metaphysics* V, 6, 1016b6–9).

We have just an accidental unity—'a mere heap' (*Metaphysics* VIII, 6, 1045a 9)—when, for example, we randomly put papers one on top of the other, and, as a result, we get a pile of papers. Every single paper preserves its substantiality; their unity only adds an accidental property to them—being above or below another paper.

Finally, we have a unity of order when the parts preserve their substantiality and their ends, but their order also enables them to contribute towards a common end. The pursuit of an end calls for a specific order of the parts. Aristotle notes that substances cannot consist of substances (*Metaphysics* 1039a3–4, 1041a4– 5). As social collectives are made up of individual substances, they cannot be substantial wholes. The system's unity stems from the orientation towards an end, which implies an order. This order has a normative nature: if we want to attain a goal, we have to act in a specific way (and we need to have authority), and this implies several stipulations and conditions, depending on the particular purpose of a specific social collective.

For Aristotle, having a common cause is the condition of all the wholes that are not a mere aggregation of parts: 'In the case of all things which have several parts and in which the totality is not, as it were, a mere heap [*soros*], but the whole is something besides the parts, there is a cause' (*Metaphysics* VIII, 6, 1045a 9–10). For social collectives, the cause is a final cause that imposes an order on the parts. The contemporary systemic approach also features a teleological tenet: the end is the unifying core of wholes (von Bertalanffy, [1968] 1973, pp. 14, 44ff.), and human and social realities pertain to the realm of open systems (von Bertalanffy, [1968] 1973, pp. 38ff.).

The relation between man and machine is also systemic, and only by dealing with it systemically can we take into account the nature of human beings, which establishes ethical conditions that are not external but intrinsic to the system. This is particularly relevant today because contemporary times are characterized by an unprecedented global technological development (von Bertalanffy, [1968] 1973, p. 215).

There is a sizeable literature on the system of human-machine interaction and on the human-computer interaction (HCI). The concept of machine builds on the notion of artifact, as it performs a task. Apart from specific technical matters, current literature points out that human beings must come first, and it sets forth several criteria to meet this goal.[3] Erich Fromm talks about connecting the 'system man' with the whole system, so that technologies—and specifically computers—serve people's nature, values, and growth (1968, pp. 4, 97ff.).

Concerning AI, recent literature conceives the relation human-AI as a teammates' relationship.[4] The key to a successful team connection is that all members share the same goals. ChatGPT, for example, aims to correctly inform users about the prompts received. With its goal determined by its human designers, ChatGPT learns from the information online and from the feedback provided by users. Referring to ChatGPT, Mark Coeckelbergh and David Gunkel remark that 'humans and technologies are entangled with one another' (2024, p. 2223). For these authors, truth and meaning are co-created by humans and nonhumans—they are socially and technologically generated (2024, p. 2228). Waze has been designed to guide drivers along the best route to get to a destination, feeding on information provided by host of drivers to achieve this goal. Thus, a particular system is created with AI and humans to seek the common ends established by human designers. The team breaks down if members' goal diverge—for instance, if an individual feeds fake news to ChatGPT.

What is the ontology of this AI-humans system? The system has an accidental unity stemming from the common end sought. This is not automatic; it involves a normative task that needs to be performed. Hence, this unity has to be maintained, which means adapting to the requirements of a respectful attitude towards human beings, and so it is a dynamic system. In contrast, Byung-Chul Han (2017, pp. 10–11) characterizes what he calls the 'digital swarm' as comprising isolated individuals without a 'soul' to unite them, without harmony—if this is the case, we are walking on a misleading path.

[3] See, for example, Janet C. Read (2024). Tim Gorichanaz (2022) applies virtue ethics to the HCI.

[4] See Anna C. Trujillo et al. (2019).

The Ontology of Artifacts and AI

To establish whether AI agents can think and be moral or not (the topics of upcoming chapters), the first step is to ascertain the meaning of the 'A' in AI —that is, to establish what artifacts are.[5] Though 'riddled with more questions than answers' in its analysis of the definition of artifacts, the *Stanford Encyclopedia of Philosophy* starts the corresponding entry—'Artifact' (Beth Preston, 2022)—by citing Risto Hilpinen's (1992) definition: objects made intentionally in order to accomplish some purpose. Thus, there are three conditions that artifacts must meet: having been produced intentionally, involving a modification of materials, and having a purpose. More recent literature on social ontology delves into the social nature of artifacts: in some cases, their nature does not only depend on their creators' intentions, but also on their collective acceptance by specific social groups. Their nature may even depend exclusively on their collective acceptance (see Tim Juvshik, 2024 for an elaboration on the many possibilities at play). This approach favours the idea of viewing AI as a system.

Preston (2022) explains the primary tenets regarding the existence and nature of artifacts. The notion of more or less dependence on intentionality is common to almost all the theories she described. However, an interesting remark made by Steven Vogel (referring to environmental effects, but applicable to other scenarios, particularly to forms of AI) should be considered. He states, 'To construct an artifact is always to put processes into motion whose workings we do not fully understand and whose consequences we do not fully foresee' (2015, p. 112).[6]

David Wiggins (not mentioned by Preston) begins his book *Sameness and Substance Renewed* (2001) on the identity of entities by claiming that 'the chief aim of this book is to elaborate a theory of the individuation of continuants, including living substances and other substances' (2001,

[5] In AI the AI strictly means 'artificial'. According to the Merriam-Webster Dictionary 'artificial' is the adjective ('made, produced, or done by humans') and 'artifact' is the corresponding noun ('a product of artificial character').

[6] As early as 1960, Norbert Wiener warned that machines may prove to 'be both effective and dangerous' (1960, p. 1355). In my opinion, this sensible take on machines is linked to the current discussion on authorship regarding the results of ChatGPT elaborations. See Coeckelbergh and Gunkel (2024), van Woudenber et al. (2024) and Irmak (2024). Artificial entities may exceed the expectations of their designers, who do not create things ex-nihilo and 'play with' poorly/incompletely known/understood. Still, these entities cannot exceed the limitations of their artificial nature.

p. 1). These 'other substances' include artifacts. In fact, he makes a distinction between natural substances and artifacts,[7] pointing out:

> Artefacts are things to be encountered in nature, and subject to its laws. But they need not be collected and classified together as this or that artefact in virtue of internal resemblances or scientific resemblances. Artefacts are collected up not by reference to a theoretically hypothesized common constitution but under functional descriptions that are precisely indifferent to specific constitution and particular mode of interaction with the environment. (Wiggins, 2001, p. 87)

While natural substances are classified by their nature, artifacts are classified by their function. There may be many kinds of clocks; what they have in common is that they measure time: 'A clock is any time-keeping device; a pen is any rigid ink-applying writing implement, and so on' (2001, p. 87). This functional description, Wiggins asserts, specifies a nominal feature, not a real essence (2001, p. 87). The key difference is that natural substances, the author states, have natural propensities stemming from their essence and constituting their foundation (2001, p. 88), while artifacts do not have them. Wiggins had quoted Aristotle on this point before (2001, p. 81) and cited him again concerning this distinction between natural substances and artifacts:

> Of things that exist, some exist by nature, some from other causes. 'By nature' the animals and their parts exist, and the plants and the simple bodies (earth, fire, air, water)—for we say that these and the like exist 'by nature.' All the things mentioned present a feature in which they differ from things which are not constituted by nature. Each of them has within itself a principle of motion and of stationariness (in respect of place, or of growth and decrease, or by way of alteration). On the other hand, a bed and a coat and anything else of that sort, qua receiving these designations i.e. in so far as they are products of art—have no innate impulse to change. But in so far as they happen to be composed of stone or of earth or of a mixture of the two, they do have such an impulse, and just to that extent which seems to indicate that nature is a source or cause of being moved and of being at rest in that to which it belongs primarily, in virtue of itself and not in virtue of a concomitant attribute. I say 'not in virtue of a concomitant attribute,' because (for instance) a man who is a doctor might cure himself. Nevertheless it is not in so far as he is a patient that he possesses the art of medicine: it

[7] So does Kenneth Himma (2009, p. 21).

merely has happened that the same man is doctor and patient—and that is why these attributes are not always found together. So it is with all other artificial products. None of them has in itself the source of its own production. But while in some cases (for instance houses and the other products of manual labour) that principle is in something else external to the thing, in others those which may cause a change in themselves in virtue of a concomitant attribute – it lies in the things themselves (but not in virtue of what they are). (*Physics* II, 1, 192b8–32)

The internal adequate disposition of the natural parts of artifacts enables them to operate. A related passage is found in *Nicomachean Ethics* (VI, 4, 1140a11–16):

All art is concerned with coming into being, i.e. with contriving and considering how something may come into being which is capable of either being or not being, and whose origin is in the maker and not in the thing made; for art is concerned neither with things that are, or come into being, by necessity, nor with things that do so in accordance with nature (since these have their origin in themselves).

Aristotle believes that, while natural substances have an internal action principle or power, artifacts get this action principle from their producers as a result of an appropriate combination of natural components.[8] James Jacobs (2019, p. 158) explains, 'an artifact is not simply any aggregate of independent substances, but is rather an ordered aggregate that allows man to make use of the natural powers of the substances involved.'

Furthermore, natural substances, due to their inner action principle, lean towards an end. As Charlotte Witt remarks, 'Aristotle's dispositional definition of causal powers rests upon a teleological footing' (2008,

[8] Lynne Rudder Baker (2004, p. 105) thinks that artifacts also have an internal action principle, but she does not make a distinction between natural and artificial objects (2004, p. 102). She is part of the supporters of 'function essentialism' (for more on this position, see Tim Juvshik 2021). However, the fact that artifacts cannot operate if they are not built and fuelled implies that their action principle lies outside of them. For her, artifacts are dependent on builders' or users' intentions, which implies a distinction from natural substances. Kathrin Koslicki (2021) applies the hylomorphic theory to artifacts, stating, 'Artifact-forms, according to the present conception, are functions, i.e., actual or potential capacities, that are located in the objects themselves, and not in the minds of agents who make or use artifacts' (2021, p. 118). Artifacts have a specific function—an extrinsic capacity—due to their form/structure and the intrinsic capacities of the materials that they are made of.

p. 130). Aristotle notes, 'For those things are natural which, by a continuous movement originated from an internal principle, arrive at some completion: the same completion is not reached from every principle; nor any chance completion, but always the tendency in each is towards the same end, if there is no impediment' (*Physics* II 8 199b15). Then, according to Aristotle, there is a natural end for every natural substance: 'nature does nothing in vain' (*On the Gait of Animals* 2, 704b12–17), and 'nature belongs to the class of causes which act for the sake of something' (*Physics* II, 8, 198b10–11). Instead, in the case of artificial entities, this end is infused by artifact creators and users. Jacobs posits:

> […] the craftsman cannot create a real substance with a real essence; rather, he uses already existing substances as instruments to an end, and so the artifact does not have real existential unity. It has an intentional unity arising from that conceived form accidentally uniting those substances; it is this intentionality that is the extrinsic source of unity for the artifact and so acts as the essence of the artifact. (2019, p. 153)

With their free will, human beings can act (or not) according to their natural end, and the realm of morality stems from this very fact.[9] Instead, as artifacts have neither an internal action principle tending to an end nor free will, their end is infused in them by human beings, who are responsible for their morality. The 'accidental unity of order' also applies here, as impressed by the designer with a view to the system's purpose.

Nonetheless, there is an objection to Aristotle's division between natural beings and artifacts. A 'continuum problem' emerges when dealing with both, and a growing body of literature delves into this issue (see Preston, 2022). Some artifacts are created unintentionally (for example, paths created by people repeatedly going from one place to another), and some natural things are used as artifacts (for example a stone used as paperweight). There are also 'artifacts' produced by animals such as bird nests, spider webs, and beaver dams. Therefore, the natural-artificial divide does not seem to be all that clear. Hilpinen speaks about 'naturefacts,' pointing out, 'Objects taken from their natural environment and used as tools or for some other purpose form a bridge between natural objects and artifacts […]. When a naturefact is modified to improve its usability for some purpose, it becomes a genuine artifact' (2018, Section 1). For Maria

[9] I will circle back to this topic in Chap. 8.

Elizabeth Reicher, instead, the decisive criterion is not a modification, but the author's intentionality to consider the naturefact as a type of artifact (cf. 2022, p. 244).

Aristotle has dealt with *téchne* in such a way that it is not affected by the 'continuum problem' (see Errol Katayama, 2024). He refers to animals' products in *Physics* II, 8, 199a23–24: 'they make things neither by art [*téchne*] nor after inquiry or deliberation.' As explained in Chap. 2, *téchne* goes beyond experience, which is something that some animals may have—but experience is all they have. Similarly, the stone used as paperweight is not an artifact, because it has not been produced as a result of technical knowledge: the stone has become a paperweight simply because of its owner's choice. An unintentionally formed path is not an artifact because there was no intentional design involved in its creation.

What kind of artifact is AI? Though it needs matter to exist (a computer, energy, etc.), it is not matter itself. Nurbay Irmak has argued that words (2019), musical works, and software (2012) are abstract artifacts, which do not have a spatial location but are created by human beings for an end. He describes them as 'temporal, nonspatial, repeatable, contingent entities that exist in virtue of certain purposeful human activities' (2019, p. 1147). Reicher (2022) also talks about abstract artifacts, claiming that they have primacy over material artifacts. She believes that their supremacy is temporary (because first we think or design the artifact as an 'artifact type,' and then we manufacture it), ontological (in the sense that concrete artifacts are ontologically dependent on abstract artifacts), and finally conceptual.

Irmak also characterises algorithms (sequence of instructions to accomplish a certain goal) as abstract artifacts—a view shared by Subrata Dasgupta (2016, p. 60). I think that we can apply this characterisation to AI as well,[10] for it is indeed a set of algorithms. In any case, the relevant issue from an ontological point of view is that AI is an artifact.

Yet, there is another characteristic of AI that is especially relevant: the fact that it can work autonomously based on algorithmic criteria and inferences from the information provided. Therein lies the aforementioned possibility grasped by Norbert Wiener when he noted, 'It is my thesis that machines can and do transcend some of the limitations of their designers,

[10] Reicher also shares this view, as she wrote in a personal correspondence (7 January 2025), 'I agree with you: Software products (and this is what AI consists in, isn't it?) are abstract artifacts.'

and that in doing so they may be both effective and dangerous' (1960, p. 1355), and underscored by Vogel's remark quoted earlier and highlighting that they may have 'consequences we do not fully foresee' (2015, p. 112). Aristotle had also considered that some artifacts 'may cause a change in themselves,' as the last part of the above quoted long paragraph states.

Kaya Ismael (2018) states that AI 'is a group of algorithms that can modify its algorithms and create new algorithms in response to learned inputs and data as opposed to relying solely on the inputs it was designed to recognize as triggers. This ability to change, adapt and grow based on new data, is described as "intelligence".'

According to Davide Panagia, algorithms are 'automated systems for the perpetual arrangement and rearrangement of relations,' which means that 'they barter in probabilities and correlations and not verifiable or derivable truths' (2021, pp. 2 and 7). 'The algorithm's dispositional power is a power of perpetual arrangement and rearrangement of data that measure weights and values which in turn generate world relations that are neither indexical nor analogue, but virtual' (2021, pp. 9–10). Virtuality and negative feedback are the algorithm dispositif's modes of existence, with the latter being necessary to match AI's behaviour with its goal.

Louise Amoore considers 'the algorithm not as a finite series of programmable steps but as perennially adjustable and modifiable in relation to a target output' (2020, p. 13). She 'understand[s] the spatial logic of algorithms to be an arrangement of propositions that significantly generates what matters in the world' (2020, p. 12) and that is both transformative and performative (2020, p. 13). In turn, Rob Kitchin points out that algorithms are ontogenetic in nature—'always in a state of becoming' (2017, p. 18), somewhat uncertain, relational, contingent, performative, not neutral.

This autonomous nature of AI involves matters of ethical, political, and social responsibility, because, like all technical artifacts, it must serve the human good, as argued in Chap. 2. I will revisit this topic in Chaps. 6, 7, 8, and 9.

Conclusion

In this chapter, I have explored the characteristics of the AI system. First, I explained what a system is and elaborated on its characteristics. Next, I described the man-machine system and the man-AI system, drawing on

Aristotle's ideas to establish the ontological nature of these systems as ordered wholes guided by common final causes. Second, I set out to research the nature of AI, which I consider an artifact, as the 'A' in its acronym indicates. I also relied on Aristotle's notions to explore the nature of artifacts and forge a definition: they are accidental wholes composed of natural elements. They differ from natural things because they do not have an internal dynamic principle, and their end is provided by their designers and users. Finally, I characterised AI as an abstract artifact with a very relevant and impactful feature: autonomy.

References

Amoore, L. (2020). *Cloud Ethics. Algorithms and the Attributes of Ourselves and Others*. Duke University Press.

Aristotle. *Nicomachean Ethics* (W. D. Ross, Trans.). Retrieved October 20, 2024, from https://classics.mit.edu/Aristotle/nicomachaen.html.

Aristotle. *On the Gait of Animals* (A. S. L. Farquharson, Trans.). Retrieved October 20, 2024, from https://classics.mit.edu/Aristotle/gait_anim.html

Aristotle. *Physics* (R. P. Hardie and R. K. Gaye, Trans.). Retrieved October 20, 2024, from https://classics.mit.edu/Aristotle/physics.html.

Coeckelbergh, M., & Gunkel, D. J. (2024). ChatGPT: Deconstructing the Debate and Moving it Forward. *AI & Society, 39*, 2221–2231.

Dasgupta, S. (2016). *Computer Science: A Very Short Introduction*. Oxford University Press.

Fromm, E. (1968). *The Revolution of Hope. Towards a Humanized Technology*. Bantan Books.

Gorichanaz, T. (2022). Designing a Future Worth Wanting: Applying Virtue Ethics to HCI. https://arxiv.org/abs/2204.02237. Retrieved November 20, 2024, from https://doi.org/10.48550/arXiv.2204.02237

Han, B.-C. (2017). *The Swarm*. The MIT Press.

Hilpinen, R. (2018). Artifact. In E. N. Zalta (ed.), *The Stanford Encyclopedia of Philosophy (Summer 2018 Edition)*. Retrieved October 20, 2024, from https://plato.stanford.edu/archives/sum2018/entries/artifact/.

Hilpinen, R. (1992). Artifacts and Works of Art. *Theoria, 58*(1), 58–82. https://doi.org/10.1111/j.1755-2567.1992.tb01155.x

Himma, K. E. (2009). Artificial Agency, Consciousness, and the Criteria for Moral Agency: What Properties Must an Artificial Agent Have to be Moral Agent? *Ethics and Information Technology, 11*, 19–29.

Irmak, N. (2012). Software is an Abstract Artifact. *Grazer Philosophische Studien, 86*, 55–72.

Irmak, N. (2019). An Ontology of Words. *Erkenntnis, 84*, 1139–1158.

Irmak, N. (2024). Artifacts Without Authors: Generative Artificial Intelligence and the Question of Authorship. *Metaphysics, 7*(1), 1–15.

Ismael, K. (2018, October 26). AI vs. Algorithms: What's the Difference? *Information Management*. Retrieved February 2, 2024, from https://www.cmswire.com/information-management/ai-vs-algorithms-whats-the-difference/.

Jacobs, J. M. (2019). The Ontological Status of Artifacts. In C. Cullen & F. Harkins (Eds.), *The Discovery of Being & Thomas Aquinas: Philosophical and Theological Perspectives* (pp. 145–160). The Catholic University of America Press.

Juvshik, T. (2021). Function essentialism about artifacts. *Philosophical Studies, 178*(9), 2943–2964.

Juvshik, T. (2024). On the Social Nature of Artifacts. *Theoria, 89*(6), 910–932.

Katayama, E. G. (2024). Aristotle on Artificial Products. In D. Keyt & C. Shields (Eds.), *Principles and Praxis in Ancient Greek Philosophy: Essays in Ancient Greek Philosophy in Honor of Fred D. Miller, Jr* (pp. 227–249). Springer Verlag. https://doi.org/10.1007/978-3-031-51146-2_12

Kitchin, R. (2017). Thinking Critically About and Researching Algorithms. *Information, Communication and Society, 20*(1), 14–29.

Koslicki, K. (2021). Form, Matter, Substance. Inaugural Lecture, *Chronique Universitaires 2020–2021*, pp. 99–119. Université de Neuchâtel.

Panagia, D. (2021). On the Possibilities of a Political Theory of Algorithms. *Political Theory, 49*(1), 109–133. Retrieved 20 July 2024, from https://journals.sagepub.com/doi/epub/10.1177/0090591720959853

Preston, B. (2022). Artifact. In E. N. Zalta & U. Nodelman (eds.), *The Stanford Encyclopedia of Philosophy* (Winter 2022 Edition). Retrieved January 12, 2025, from https://plato.stanford.edu/archives/win2022/entries/artifact/.

Read, J. C. (2024). Ethics in HCI. In C. Stephandis & G. Salvendy (Eds.), *Foundations and Fundamentals in Human-Computer Interaction* (pp. 429–448). CRC Press. Taylor & Francis Group.

Reicher, M. E. (2022). The Primacy of Abstract Artifacts. In P. McNamara, A. Jones, & M. Brown (Eds.), *Agency, Norms, Inquiry, and Artifacts: Essays in Honor of Risto Hilpinen* (Synthese Library) (Vol. 454). Springer.

Rudder Baker, L. (2004). The Ontology of Artifacts. *Philosophical Explorations, 7*(2), 99–111. https://doi.org/10.1080/1386979041000l694462

Searle, J. R. (1994). *The Rediscovery of the Mind*. The MIT Press.

Trujillo, A. C., Gregory, I. M., & Ackerman, K. A. (2019). Evolving Relationship between Humans and Machines. IFAC PapersONLine 51–34, pp. 366–371. Retrieved October 20, 2024, from https://www.researchgate.net/publication/331007498_Evolving_Relationship_between_Humans_and_Machines.

Van Woudenberg, R., Ranalli, C., & Bracker, D. (2024). Authorship and ChatGPT: A Conservative View. *Philosophy & Technology, 37*, 34.

Vogel, S. (2015). *Thinking Like a Mall.* The MIT Press.
Von Bertalanffy, L. ([1968] 1973). *General System Theory.* Penguin University Books.
Wiener, N. (1960, May 6). Some Moral and Technical Consequences of Automation. *Science, New Series, 131*(3410), 1355–1358.
Wiggins, D. (2001). *Sameness and Substance Renewed.* Cambridge University Press.
Witt, C. (2008). Aristotelian Powers. In R. Groff (Ed.), *Revitalizing Causality* (pp. 129–138). Routledge.

CHAPTER 5

Does AI Really Think?

Abstract This chapter seeks to answer the question in its title. After trying to ascertain what intelligence is, it introduces the Aristotelian notion of human intelligence and its process, before analysing if AI can follow this process, only to conclude that AI's process differs in its very essence from Aristotle's intelligence concept. Abstraction and abduction are exclusively human activities, and human intelligence grows in the whole context of human beings, with their sensorial, emotional, moral, spiritual, social, and environmental components. To call AI 'intelligent' is an anthropomorphising misrepresentation.

Keywords Intelligence • Thinking • AI Intelligence • AI Thinking

While the last chapter focused on the 'A' of AI, this chapter will zero in on its 'I.' As Evandro Agazzi states, the answer to the question 'Can machines think?' concentrates on the notion of thinking. He phrases this query differently, 'Can an automaton equal man's thinking activity?' (1981, p. 220). The answer implies knowing the characteristics of human thinking—which hinges on the nature of human beings—and analysing if AI shares those features.

R. F. Crespo, *A Classical Philosophical Approach to Artificial Intelligence*, https://doi.org/10.1007/978-3-032-18462-7_5

Mario Bunge has previously entitled an article 'Do Computers Think?' ([1956] 1959). Bunge's argument is grounded on the fact that 'Unlike nature and unlike artifacts, man is able to build a mental level upon the physiological one' (Bunge [1956] 1959, p. 131). Whether it is clear or not where this mental level comes from, the core of this statement is that man has a mental—not physical—level, which artifacts lack. This mental level lies at the root of human beings' consciousness, awareness, and their abstracting capacity (Bunge [1956] 1959, pp. 137, 138, 141). Instead, 'computing machines just perform certain operations without being aware of it: they do not know what they are doing or even *that* they are doing anything—and this, simply because machines have no consciousness' (Bunge [1956] 1959), 136).[1] Indeed, this primary question also kicks off Alan Turing's, 1950 article, 'Can Machines Think?' (Turing, 1950, p. 433).

But what is to think or to know? Knowledge is a wider notion than thinking; it is a result of perceiving and thinking. Thinking goes beyond the mere recording of data and the logical processes that can be carried out with them. What is the surplus of thinking? This has been discussed at length since the time of classical Greek philosophy. For Aristotle, the purpose of thinking is capturing the essence of things as well as the causes of events that are imprinted in our souls as a result of an intentional process.

These questions explore the possibility of machines doing something that resembles human thinking. The goal to at least mimic or simulate human thinking is implicitly and explicitly included in most definitions of AI. According to *Encyclopedia Britannica*, for example, AI is:

> the ability of a digital computer or computer-controlled robot to perform tasks commonly associated with intelligent beings. The term is frequently applied to the project of developing systems endowed with the intellectual processes characteristic of humans, such as the ability to reason, discover meaning, generalize, or learn from past experience (https://www.britannica.com/technology/artificial-intelligence)

However, though the previous questions by Agazzi, Bunge, and Turing ask about thinking, the word used in the expression AI and in this

[1] For an excellent essay on the mental—not physical—condition of consciousness, see Thomas Tozer (2020).

Encyclopedia Britannica definition is not 'thinking' but 'Intelligence.' Is intelligence equal, different, or partially equivalent to thinking?

A widely quoted article by Linda Gottfredson (1997, p. 13) describes intelligence as 'a very general mental capability that, among other things, involves the ability to reason, plan, solve problems, think abstractly, comprehend complex ideas, learn quickly and learn from experience.' She uses the expressions 'comprehending,' 'catching on,' and 'making sense of things' to refer to intelligence. A report by Shane Legg and Marcus Hutter (2007) contains 'A Collection of [70] Definitions of Intelligence.' The authors have classified these definitions as 'collective' (proposed by groups, and organizations including dictionaries), psychologist, and AI researcher definitions. In the 18 definitions included in the first group, the words 'knowledge,' 'abstract knowledge,' 'understanding,' 'comprehend,' and 'thinking' appear in almost all. The 35 psychologist definitions refer to the technical function of intelligence, although the expression 'abstract thinking' also shows up. In the 18 AI researchers' definitions, the words 'thinking,' 'knowledge,' and 'understanding' are missing altogether.

Thus, AI has the 'intention' to do what humans do when using their intelligence, but, at the same time, it does away with the functions that broad definitions of intelligence include. Still, the AI reference to human intellectual processes implicitly relates to those functions. Does this relation suffice to characterise AI as truly intelligent?

Aristotle classified terms with related but different meanings. In book I of *Categories* (1a), he made a distinction between equivocally (*homónyma*) and univocally (*synónyma*) named things, but he also considered other possibilities. Christopher Shields (2002) has extensively studied this topic, finding two kinds of equivocity or homonymy in Aristotle: discrete—when meanings differ completely—and comprehensive—when meanings partially overlap.[2] An example used by Aristotle to illustrate the latter is 'healthy': this word has a focal meaning, which refers to a human body in good health, and derivative meanings pointing to salubrious foods, sports, medicines, and so on (cf. *Metaphysics*, IV, 2, 1003a32 and ff.). In *Topics* I, 15, Aristotle provided other examples, including, for instance, perceiving (a cognitive act) by the body and by the soul (106b24–25).

In his exploration of Aristotle's work, Shields also finds what he (2002, pp. 39–40) calls 'seductive discrete homonyms' that are non-accidental

[2] This doctrine has been developed and perfectioned by medieval authors; see E. Jennifer Ashworth and Domenic D'Ettore (2021).

discrete homonyms—that is, discrete homonyms that are not easy to spot. These are different things with the same name that are not easy to tell apart because they seem to have something in common. This might be the case of intelligence in AI. If human intelligence—what AI tries to mimic—has completely different characteristics than AI, the word 'intelligence' applied to AI would be a discrete homonymy or equivocal in respect to human intelligence.

The next steps will be first to explain Aristotle's notion of human intelligence and its process. Then, the chapter will analyse if AI can do the same process, introducing ideas posited by contemporary philosophers dealing with AI and thinking.

Aristotle on Human Intelligence

In the aforementioned collection by Legg and Hutter (2007), the most general definitions of 'intelligence' include knowledge, abstract knowledge, understanding, comprehending, and thinking as its results or activities. I am casting the other—psychological and AI researchers'—definitions aside because they entail technical meanings of the term 'intelligence' applicable to those specific fields. I view human intelligence as the process that allows people to think or know rationally. As Agazzi also affirms, '*nobody* doubts that man does think' (1981, p. 199), but there may be differing takes on the meaning and process of thinking.

For Aristotle, the capability for rational knowledge is a defining characteristic of human beings. He believed that things have necessary properties that are unique to them—they are called *idia* in Greek and are more generally known as 'propria.' Thus, a *propria* of human beings is that they are capable of rational knowledge (*Topics* I, 5, 102a18–30). In fact, Aristotle famously characterised people as the only animals who are able to use language (*zôon lógon échon*, *Politics* I, 2, 1253a9–10) to rationally express what is convenient or not, fair or unfair. In *Nicomachean Ethics,* he asserted that 'reason (*noûs*, the capacity of thinking) more than anything else is man' (X, 8, 1178a 6–7). Also referring to *noûs*, Aristotelian scholar Jonathan Lear states, 'Man is by nature a systematic understander of the world' (1988, p. 117). The human good of human beings, Aristotle noted, is their function, 'an activity of soul which follows or implies a rational principle' (*Nicomachean Ethics* I, 7, 1098a8), while human happiness hinges on *noûs* (NE X, 7).

Aristotle introduced ample descriptions of the knowledge process in *Metaphysics*, in *Posterior Analytics*, and in *De Anima*. Indeed, *Metaphysics* (I, 1, 980a20) starts with the well-known statement that 'All men by nature desire to know [*toû eidénai*].' Knowledge is prompted by a natural desire and begins with the senses, especially sight. In some animals, sensations produce the memory [*mnéme*] and, then, the experience [*empeiría*] of things. Thanks to experience, the next step is science [*epistéme*] and art [*téchne*] about the universal. Aristotle ended this first chapter of *Metaphysics* by introducing wisdom—the knowledge of principles and causes—as the most excellent form of knowing, and he delved into the characteristics of wisdom in Chap. 2.

Aristotle described what he called 'dianoethic' virtues— that is, intellectual abilities— in Book VI of his *Nicomachean Ethics* (VI, 1, 1139a1). These are the virtues of the rational part of the soul, which looks for the truth (*epistemotikón*, 1139a12), and include (*Nicomachean Ethics* VI, 3) technique (*téchne*), science (*epistéme*), practical wisdom (*phrónesis*), intuitive reason (*nous*), and philosophical wisdom (*sophia*). The latter embodies the union of intuitive reason and science, constituting scientific knowledge of the highest objects (*Nicomachean Ethics* VI, 7), while technique and practical wisdom are dispositions towards action. All theoretical knowledge, Aristotle noted referring to science, 'proceeds sometimes through induction (*epagogés*) and sometimes by syllogism' (*Nicomachean Ethics* VI, 3, 1139b38–39)—that is, induction and deduction, previously providing the principles of the latter by *noûs*, which captures them via induction. Science, a deductive procedure, obeys logical laws and performs its task rather automatically based on its principles. The discovery of scientific principles, as noted, involves induction [*epagogé*] and *noûs*, which are the most 'human' steps of thinking.

There are relevant passages about the role of both induction and *noûs* in *Posterior Analytics* II, 19, where Aristotle described the cognitive steps leading to the knowledge of first principles. He stated that we do not possess these principles and we would not acquire them without a preceding ability called perception, shared by animals. Perception is the first step towards the knowledge of first principles (99b 33–35). Then,

> from perception there comes memory, as we call it, and from memory (when it occurs often in connection with the same item), experience; for memories which are many in number form a single experience. And from experience, or from all the universal which has come to rest in the soul (…), there comes

> a principle of skill or understanding [*epistéme*]—of skill if it deals with how things come about, or understanding if it deals with how things are. (100a3–9)

This is *epagogé*, the inductive process,[3] but what is the state or ability that grasps the principle of skill or understanding? There is a 'jump' between experience and grasping principles that is performed by *noûs*, translated as 'intuition' by commentators and as 'comprehension' by Jonathan Barnes (comentary of Barnes in Aristotle, Posterior Analytics 1993, p. 267). The following passage translated by Barnes explains this:

> Since of the intellectual states by which we grasp truth some are always true and some admit falsehood (e.g. opinion and reasoning—whereas understanding [*epistéme*] and comprehension [*noûs*] are always true), and no kind other than comprehension is more certain than understanding, and the principles of demonstrations are more familiar, and all understanding involves an account— there will not be understanding of the principles; and since it is not possible for anything to be truer than understanding, except comprehension, there will be comprehension of the principles—both if we inquire from these facts and because demonstration is not a principle of demonstration so that understanding is not a principle of understanding either—so if we have no other true kind apart from understanding, comprehension will be the principle of understanding. And the principle will be of the principle, and as a whole will be similarly related to the whole object. (100b6–17)

Thus, thanks to induction (*epagogé*), comprehension (*noûs*) grasps the principles from which understanding (*epistéme*) demonstrates conclusions.

Noûs, translated into Latin as 'intellectus,' is the most characteristically human cognitive capability. Truly necessary to capture the essences (*De Anima* III, 430b28), *noûs* stands as the most important step of the cognitive process. Its insight draws from a rich experience, going beyond it, grasping the universal implicit in it. Dialectic reasoning, based on credible beliefs, contributes to the task of *noûs*: 'it contains the way to the first principles of all inquiries' (*Topics* 101b4). As Shields states, 'dialectic is a road to this end [*noûs*] because it helps us to systematise beliefs, to subject

[3] See C. D. C. Reeve (1995, pp. 56ff, 2006). for a complete description of this process.

them to scrutiny, and measure against other beliefs, whether empirically formed or rationally generated' (2014, p. 153).

The concepts, essences, or forms captured by *noûs* are already general notions; then, we do not need to gather a complete number of instances to know the general concept. Induction is part of this process, but Aristotle's induction (*epagogé*) should not be confused with the modern concept of enumerative induction. Jaakko Hintikka (1992, p. 34) explains the difference, 'For Aristotle, the problem of induction was not first and foremost a problem of inference from particulars to a generalisation. It was a problem of concept formation. Particular cases were stepping-stones to the concepts or forms "induced" to be realised in the soul (…) Hence there is no such problem as the justification of induction for Aristotle.' Indeed, Aristotle described induction as 'a process of inducing in ourselves the right concepts' (Hintikka, 1980, p. 429), which is made possible by *noûs*.

It may be said that, for Aristotle, the theoretical truth features two levels: a pre-propositional level of (a richer or poorer) knowledge of essences (*Metaphysics* IX, 10, 1051b17–1052a 4) and a propositional level of principles and judgments (*Metaphysics* IV, 7, 1011b25–27 and VI, 4, 1027b20–23; *De Anima* III, 6).[4] Aristotle viewed the truth not only as a logical category associated with judgments and propositions, but mainly as referring to reality as both an actual correspondence to the facts of the composition or division of terms expressed by judgments, and the ability to grasp the actual beings that these terms designate, which become present in the souls. 'Actual knowledge is identical with its objects' (*De Anima* III, 7, 431a1), so that knowledge is an 'intentional' possession of the form of what is known. This form is a universal concept abstracted from the particulars as a result of the work carried out by *epagogé* and *noûs*.

Aristotle's *De Anima* III, 4–8 deals with knowledge, especially with *noûs*, stressing that *noûs* and perception are the two parts of the soul that produce the immaterial presence of the essence of things known in it (III, 4, 429a13–19). The 'thinking soul' houses the potential of these forms [*tà eíde*] (III, 4, 429a27–28 and 8, 431b24–29). Before thinking and different from perception, Aristotle introduced *phantasia* (image, III, 7, 431a14–16), noting that the soul never thinks without it. Images [*phantasmata*] remain in the soul after perception, and *noûs* thinks the forms in *phantasmata* (III, 7, 431b2). As Corcilius et al. (2024, p. 44) point out,

[4] Cf. Alejandro Vigo (2006, pp. 137ff).

phantasia is not a cognitive ability but an enabling condition of the cognitive state beyond perception.

Noûs provides the essences of things and principles of science. Referring to science (*epistéme*), as mentioned earlier, Aristotle elaborated that all theoretical knowledge 'proceeds sometimes through induction (*epagogés*) and sometimes by syllogism' (*Nicomachean Ethics* VI, 3, 1139b28–29). For him, induction captures the universal through *noûs* (39–40), while deduction derives from the universal.

Finally, another Aristotelian notion should be taken into account. American philosopher Charles Peirce underscored that, in *Prior Analytics* II, 25, Aristotle introduced a form of induction that he called *apagogè*. In any syllogism, this imperfect reasoning infers the minor premise from its major premise and its conclusion—as in, for example,

All virtues are good.
Justice is good.
Justice is a virtue.

This syllogism is invalid because we cannot infer that justice is a virtue from the fact that justice is good, but this inference is indeed probable.

Peirce referred to this inference as 'abduction,' considering it—in a secondary, broadened meaning—a means to arrive at scientific hypotheses. He formulated it as follows (1958 CP 5.189):

- The surprising fact *C* is observed.
- But if *A* were true, *C* would be a matter of course.
- Hence, there is reason to suspect that *A* is true.

The concept implied in this formulation is more general than the one in the syllogism above. 'A' might be a case or a hypothetical rule (see Niiniluoto, 1999, p. S439), but Peirce stated that, given that the fallacy of affirming the consequent remains, this hypothesis discovery and postulation procedure is only a first step in scientific research. That is, abduction—in this second meaning—is a heuristic method assisted by some criteria formulated by Peirce, which serve as antecedents for today's so-called epistemic virtues, like simplicity (or parsimony), unification power, fertility, and testability.

It is clear, then, that this second sense of abduction should be used to suggest hypotheses, not to accept them. Nevertheless, Niiniluoto (1999,

p. S442) argues that Pierce considered 'an extreme case of abductive inferences' that are 'irresistible or compelling' and come to us 'like a flash.' In cases like that one, Niiniluoto states, 'for Peirce [abduction] is not only a method of discovery but also a fallible way of *justifying* an explanation' (ibid.). In other words, the strength of this flash would produce a change of the agent's epistemic state, meaning that it amounts to more than statistical combinations.

Where does this flash come from? It is a matter of creativity that has to do with background knowledge and the work of *noûs*. Among the numerous possible hypotheses that scientists or ordinary people can think of, they choose one that they regard as more plausible—and they often get it right. This why Peirce spoke about a guess instinct (1958 CP 6.476, 1908; CP 7.220, 1901).[5]

In sum, Aristotle developed a theory of knowledge that, starting from external senses and moving through memory, experience, *phantasía*, *epagogé*, *noûs*, and deduction, produces science (*epistéme*). The ability to think (*noûs*), capturing essences and causes, lies at the core of this process. Abduction, as formulated by Peirce, provides the hypotheses that will be then proved (or not) empirically. Thus, the time has come to analyse what AI can do.

AI's Actual Thinking

In a nutshell, AI can mimic enumerative inductions and deductions, and it can recognize patterns, but it cannot conduct an abstractive induction from an intuition of essences and causes.

Concerning deduction, AI's problem is that it only uses logical rules to analyse propositions, without understanding their meaning and relevance (see Erik Larson, 2021, pp. 107–115). Evandro Agazzi (1981, pp. 212–214) notes that formal deduction is only a surrogate of our logical demonstration process. The intuitive link of logical consequence based on the notion of truth is replaced by the relation between deduction and its rules. AI remains at this latter level, unable to judge its formal deductions, because it has no semantic power.

Unlike Aristotle's *epagogé*, AI's induction is enumerative and, therefore, always fallible, for we can very seldom consider all possible cases, and we do not know the future. Induction certainly adds to the process of

[5] For more on Peirce's 'guess instinct,' see, for example, Sami Paavola (2005).

knowledge, suggesting possible hypotheses, but they must be proved. As Agazzi puts it, 'all automata are able to do is to make generalisations of a probabilistic and statistical kind' (1981, p. 217). In addition, Jobst Landgrebe and Barry Smith (2023, p. 17) state that even 'Machines do not engage in inductive reasoning; they rather compute local minima for loss functions, which can be seen as a very primitive emulation of induction.'

According to Byung-Chul Han (2022, pp. 41–42), AI can only achieve the state of data correlation, failing to know why they correlate, ignoring the cause of correlation and the idea connecting data—and thus nothing is *understood* by AI. AI receives information, but it does not gain knowledge (2022, p. 7). This is why Bender et al. (2021) call 'Language Models' like GPT 'stochastic parrots': there is no reference to meaning in them (2021, p. 617). As Hamid Ekbia et al. (2015, p. 9) remark, big data 'save the appearance but not necessarily the phenomenon.'

Neither does AI have the abduction ability that involves creativity and a plethora of information about the real world (see Erik Larson, 2021, Chapter 12, Mariana Olezza, 2025 and Bertolaso & Marcos, 2023, p. 51).[6] Han describes this shortcoming very nicely: AI lacks the affective essential to human knowledge—it is not preceded by a familiarity with the entire world, and it lacks the *pathos* that prompts thinking—and, hence, it only computes (cf. 2022, pp. 37–44). Indeed, there is a tacit knowledge that AI does not possess, and Michael Polanyi explores the relevance of this kind of knowledge in his book *The Tacit Dimension* ([1966] 2009). Indeed, according to Aristotle, the mind is not separated from the body: the mind is an embodied mind. For him, the *psyche* (soul) 'is the first actuality of a natural organic body' (*On the Soul* II, 1, 412b 4). He also argues that 'still to say that the soul gets angry is as if one were to say that the soul weaves or builds a house. Probably it is better not to say that the soul pities, or learns, or thinks, but to say rather that the soul is the instrument whereby man does these things' (*On the Soul* I, 4, 408b 13–15). The human mind does not exist without a body. While there are efforts

[6] The theses of the 'extended mind' and the 'scaffolding mind' may also be used to make the case for the additional value of human knowledge versus AI (see Julian Kiverstein et al. (2013).

underway to build robots with human-like bodies, AI will never possess a body like ours; it will never have our comprehensive experience.[7]

The recent announcement of a groundbreaking AI art auction at Christie's explains why 'Contrary to how it's often described, AI is inherently human. 'Code can be considered a type of craft,' says Sebastian Sanchez, Christie's Manager of Digital Art. There are always hands involved in some shape or form, although not necessarily in the traditional sense as with painting or craftsmanship. Engineers program networks, and some artists use pre-existing tools, generating art through refined text prompts (What Is AI Art? | Christie's).[8] Unlike human beings, AI does not enjoy the intellectual pleasure that comes from understanding or creating something.

Jorge Luis Borges' renowned short story 'Funes the Memorious' can help us to understand the boundaries constraining AI.[9] The author describes Ireneo Funes, a man with a prodigious memory. He wrote:

> We, at one glance, can perceive three glasses on a table; Funes, all the leaves and tendrils and fruit that make up a grape vine. He knew by heart the forms of the southern clouds at dawn on the 30th of April, 1882, and could compare them in his memory with the mottled streaks on a book in Spanish binding he had only seen once and with the outlines of the foam raised by an oar in the Rio Negro the night before the Quebracho uprising. These memories were not simple ones; each visual image was linked to muscular sensations, thermal sensations, etc. He could reconstruct all his dreams, all his half-dreams. Two or three times he had reconstructed a whole day; he never hesitated, but each reconstruction had required a whole day. He told me: "I alone have more memories than all mankind has probably had since the world has been the world.

[7] See, e.g., Louise Barrett and Dietrich Stout (2025), Andrzej Porebski and Jakub Figura (2025), Hugo L. Rufiner et al. (2025).

[8] Tozer (2020, p. 107) explains that AI's artistic creations are entirely artificial and dependent upon the creativity of the humans whose work is fed into it.

[9] 'Funes, el memorioso' ('Funes the Memorious') was first published in page three of the Arts and Letters section of *La Nación*, an Argentine newspaper, in Buenos Aires, on June 7, 1942. English translation: https://vigeland.caltech.edu/ist4/lectures/funes%20borges.pdf. I think that the same idea is conveyed by T. S. Eliot in this line from his play *The Rock*, 'Where is the knowledge we have lost in information?' (*The Waste Land and Other Poems*, Harcourt Brace and Co., New York and London, 1934, p. 81).

Like AI, Funes was able to store an immense amount of information and to connect data. Borges later added:

> With no effort, he had learned English, French, Portuguese and Latin. I suspect, however, that he was not very capable of thought. To think is to forget differences, generalize, make abstractions. In the teeming world of Funes, there were only details, almost immediate in their presence.

AI, like Funes, does not abstract—it merely gathers information. As Michael Polanyi explains, 'Comprehension can be destroyed altogether by shifting attention from its focus to its subsidiary particulars' (1958, p. 32).[10] In the same vein, Han elaborates that thinking is the ability to tell the essential from the not essential, and any excess information compromises this ability. He states, 'Thinking necessarily involves negativity: discernment, discrimination, and selection [...] A mass of information generates no truth' (2017, p. 60). In another book, Han notes that 'today, we pursue information without gaining *knowledge*. We take notice of everything without gaining any *insight*' (2022, p. 7).

AI generalizes data but does not make abstractions, failing to grasp the essence and causes—indeed, it does not look for them nor does it need them. AI captures accidents, and, when these accidents happen more than once, it is programmed to use a name to label the thing possessing these accidents, but it does not understand the meaning of this name. As stated earlier, AI only recognizes patterns and is 'agnostic' about their substance—its work is syntactic, not semantic.[11] Semantics is the human understanding of the conclusion of this process. Luciano Floridi, referring to 'large language models' like ChatGPT, remarks that 'they do not think, reason or understand [...] and they have nothing to do with the cognitive processes present in the animal world and, above all, in the human brain and mind, to manage semantic contents successfully' (2023, p. 1). AI cannot be aware of what it is doing—it can correct itself, but this is not a conscious but a 'mechanical' action. Fittingly, Shannon Vallor resorts to the mirror metaphor, stating, 'Today's most advanced AI systems are

[10] For Polanyi (1966, pp. 17–18 and *passim*), the knowledge process involves not focusing on the details ('the proximal') but 'indwelling' them ('tacit knowing') to capture a comprehensive entity ('the distal').

[11] In the previous chapter, I referred to John Searle's view that even a syntactic property does not belong inherently to a physical entity but assigned to it by the observer (Searle, 1994, Chapter 9).

constructed as immense *mirrors* of human intelligence. They do not think for themselves; instead, they generate complex reflections cast by *our* recorded thoughts, judgments, desires, needs, perceptions, expectations, and imaginings' (2024, p. 2, italics in the original).

You may or may not agree with the Aristotelian take on thinking and knowing, but most people, including philosophers and experts on AI, have concluded that the key feature of human thinking is its ability to abstract the essences and causes from particulars. Through abstraction, human beings capture or grasp the meaning of concepts, wielding a semantic power that goes beyond the mere syntactic order—and they are aware of this power. Abstraction enables humans to know what a thing is based on a limited number of cases. This core trait essentially differentiates human beings from AI, as deep neural networks need to gather a lot (millions) of cases to assign a name to a thing—and they always do it probabilistically.

Some philosophers and AI technicians recognise these AI limitations. Margaret Boden states, 'today's computers don't grasp the meaning of what they "read" or "say"' (2016, p. 39). She also provides an example, 'the Google search engine, for instance, searches terms weighted by relevance—which is assessed statistically, not semantically (that is, *without* understanding)' (2016, p. 63). The explanation she offers is that life is necessary for the mind, and the metabolism necessary for life cannot be instantiated by AI (2016, pp. 144–145).

For Melanie Mitchell (2019), computers have 'no knowledge of the meaning of these [human interpretable concepts] symbols' (2019, p. 23) and do not need them. Human beings, she argues, have two fundamental capabilities that create knowledge—abstraction and analogy (2019, pp. 242ff.)—and that AI lacks (2019, p. 246). AI also lacks 'metacognition' or 'the ability to perceive and reflect on one's own thinking' (2019, p. 260). A software may generate texts, essays, music, etc., but it does not have a real understanding of what it has generated (2019, p. 274)—i.e., it has no conscience.

Jobst Landgrebe and Barry Smith view AI as a physical, not mental reality (2023, p. 11). The human mind, they contend, cannot be mathematically modelled (2023, p. 13): human beings have goals, desires, and intentions that AI does not have (2023, p. 39). Thinking is reasoning, planning, abstracting (2023, p. 41)—'the ability to conceive, and then deliberately plan and build' (2023, p. 45). AI cannot have intentions

(2023, p. 227), experiences (2023, p. 233), semantics, and pragmatics (2023, p. 241).

As Michael Polanyi explains, 'the semantic operations attached to a formal system are functions of the mind which understands and correctly operates the system' (1952, p. 313). For Stanley Jaki (1969), only human beings are capable of abstracting and capturing universal concepts (228), while a machine, a formal system, cannot produce a truth (216). This is why AI needs the support of human beings, thus constituting an AI system. Even Alan Turing has recognised that 'One may "play about" with a machine and get the desired result, but not knowing the reason' (Blum, 2010, p. 45). He feels 'that my mind as I know it cannot be compared to a machine' (Blum, 2010, p. 51).). As Sebastian Sunday Grève (2025, p. 516) argues, 'his definitions were not meant to constitute true (metaphysical) definitions; his ultimate aim in formulating them was to enable a clear and rigorous discourse that would be conducive to the effective development of scientific theory.'

Hubert and Stuart Dreyfus (1986) assert, 'Computers are certainly more precise and more predictable than we are, but precision and predictability are not what human intelligence is about' (1986, p. xiv). For these authors, intelligence is the ability to recognise, to synthesise, to intuit. They state, 'Each of us has, and uses every day, a power of intuitive intelligence that enables us to understand, to speak, and to cope skilfully with our everyday environment [...] Analysis and intuition work together in the human mind' (1986, p. xiv). Joseph Weinzenbaum, the creator of AI development 'Eliza,' also refers to this unique ability: 'we are capable of listening with the third ear, of sensing living truth that is truth beyond any standards of provability. It is that kind of understanding, and the kind of intelligence that is derived from it, which I claim is beyond the abilities of computers to simulate' (1976, p. 222). He is not against reason, but against the imperialism of instrumental reason. Rationality should not be separated from intuition and feeling (cf. 1976, p. 256), which lead to understanding and are what AI lacks. Besides, AI does not need to intuit and feel to develop its function.

Finally, Henry Kissinger (2018) warns:

> Ultimately, the term *artificial intelligence* may be a misnomer. To be sure, these machines can solve complex, seemingly abstract problems that had previously yielded only to human cognition. But what they do uniquely is not thinking as heretofore conceived and experienced. Rather, it is

> unprecedented memorization and computation. Because of its inherent superiority in these fields, AI is likely to win any game assigned to it. But for our purposes as humans, the games are not only about winning; they are about thinking. By treating a mathematical process as if it were a thought process, and either trying to mimic that process ourselves or merely accepting the results, we are in danger of losing the capacity that has been the essence of human cognition.

In sum, for these thinkers, while AI does expand our capabilities and can perform tasks that we, human beings, cannot undertake, this does not mean that AI thinks.

Conclusion

AI collects data and transforms them into symbols that it combines, without understanding them. It stops at the symbol, without the intention to understand it semantically, as it does not need to do that in order to accomplish its purpose (Han, 2022, p. 55). AI does not think because it does not need to think (Vallor, 2024, p. 26). It has an empiricist approach in the sense that all its 'knowledge' comes from compiling, combining, and working with data.

The conclusion is that, given that AI does not think—in terms of understanding the meaning of the symbols it accumulates—the notion of intelligence that AI implies is not the same as human intelligence. Julian Nida-Rümelin and Nathalie Weidenfeld assert (2022, pp. 57–58):

> Given the striking differences between artificial and human intelligence, it should be clear that, although computers can successfully simulate thinking and even perform many human thought processes, such as algebraic operations, far more precisely and faster than humans (this already begins with the calculator), there is no underlying understanding, no problem awareness, no insight.

This is why some authors have recommended changing its name or have proposed different names for it. For example, Bertolaso and Marcos (2023, pp. 39ff.) suggest looking for a name that takes into account the fact that AI is a system, notably including human beings. Luc Julia (2019, p. 65), who has a vast track record in the AI field, suggests the name 'intelligence augmentée,' in the sense that it can add to human intelligence. We can only accept it as a technical expression of a specific field that is essentially different from human intelligence.

Weizenbaum considers that the conflation of human and artificial intelligence may have its roots in a poor conception of human intelligence, as 'merely a species of the genus "information-processing system" [...] an entirely too simplistic notion of intelligence' (1976, p. 203). Along the same lines, Han (2022, p. 43) worries about the possibility of human thinking adapting to AI and becoming mechanical. This reductive view of human thinking can be traced back to Thomas Hobbes. The recent study by the AI Research Group of the Centre for Digital Culture of the Dicastery for Culture and Education of the Holy See (Matthew J. Gaudet, Noreen Herzfeld, Paul Scherz, and Jordan J. Wales (eds.) 2024) explains:

> First, rationality and understanding become the logical manipulation of symbolically represented information; and second, intelligence becomes efficacious problem-solving (2024, p. 59). Intelligence, in the mid-to-late-twentieth century, became equated with the logical manipulation of symbolically represented information. Accomplishing this task, a properly programmed computer would be 'thinking'. (2024, p. 61)

This chapter has presented the Aristotelian notion of intelligence, which is essentially different from this simplistic take on it. Agazzi also refers to what can be regarded as a functionalist view of the human intelligence. We fall in 'the pretension of claiming that two beings are *essentially* of the same nature in they are both able to *do* a certain number of identical or very similar things (underestimating the question about *how* these two beings do the same things, and also that of the existence of *other things* which one of them may be able to do while the other may not)' (1981, pp. 196–197). Abstraction and abduction are exclusively human activities, and the human intelligence grows in the whole context of human beings, with their sensorial, emotional, moral, spiritual, social, and environmental components. To call AI 'intelligent' is nothing but an anthropomorphising mischaracterisation (see Adriana Pacani, 2024, p. 692).

Aristotle posited that human beings can access the essences and causes of things and events. AI only groups things under a common name and statistically infers a number of consequences from them. We may characterise both activities as 'intelligent'—but in the case of AI as only 'seductive discrete equivocal' intelligent.

References

Agazzi, E. (1981). Intentionality and Artificial Intelligence. *Epistemologia, IV*(Special Issue), 195–228.

Aristotle. (1954). *Nicomachean Ethics* (D. Ross, Trans. and Intro.). Oxford University Press.

Aristotle. (1957). *On the Soul* (W. S. Hett, Trans.), Loeb Classical, Harvard University Press.

Aristotle. (1958). *Politics* (E. Barker, ed. and Trans.). Oxford University Press, 1958.

Aristotle (1993) Posterior Analytics, translated with a commentary by Jonathan Barnes, Second Edition, Oxford University Press.

Aristotle. (1995). *The Complete Works of Aristotle. The Revised Oxford Translation* (J. Barnes, Ed.). Princeton University Press, 6th printing with corrections.

Ashworth, E. J., & D'Ettore, D. (2021). Medieval Theories of Analogy. In Edward N. Zalta (ed.), *The Stanford Encyclopedia of Philosophy* (Winter Edition). Retrieved August 15, 2025, from https://plato.stanford.edu/archives/win2021/entries/analogy-medieval/.

Barrett, L., & Stout, D. (2025). Minds in Movement: Embodied Cognition in the Age of Artificial Intelligence. *Philosophical Transactions Royal Society B, 379*, 20230144.

Bender, E., Gebru, T., McMillan-Major, A., & Shmitchell, S. (2021). On the Dangers of Stochastic Parrots: Can Language Models Be Too Big? In *FAccT'21, March 3–10, 2021, ACM Conference on Fairness, Accountability, and Transparency, Virtual Event, Canada*.

Bertolaso, M., & Marcos, A. (2023). *Umanesimo Tecnologico. Una riflessione filosofica sull'inteligenza artificiale*. Carocci.

Blum, P. R. (2010). Michael Polanyi: Can the Mind be Represented by a Machine? Documents of the Discussion in 1949. *Polanyiana, 201*(1–2), 35–60.

Boden, M. (2016). *AI, its Nature and Future*. Oxford University Press.

Bunge, M. ([1956] 1959). Do Computers Think? *British Journal of the Philosophy of Science, 7* (26), 139–148, reprinted in Mario Bunge (1959) *Metascientific Queries*, 124–152, Springfield, Ill.: Charles C. Thomas.

Corcilius, K., Falcon, A., & Roreitner, R. (2024). *Aristotle on the Essence of Human Thought*. Oxford University Press.

Dreyfus, H. L., & Dreyfus, S. E. (1986). *Mind over Machine. The Power of Human Intuition and Expertise in the Era of Computer*. The Free Press.

Ekbia, H., Mattioli, M., Kouper, I., & Ghazi, A. (2015). Big data, Bigger Dilemmas: A Critical Review, *Journal of the Association for Information, Science and Technology*. https://doi.org/10.1002/asi.23294. Retrieved August 3, 2024, from https://www.researchgate.net/publication/270345647_Big_Data_Bigger_Dilemmas_A_Critical_Review.

Floridi, L. (2023). AI as Agency Without Intelligence. On ChatGPT, Large Language Models, and Other Generative Models. *Philosophy and Technology, 36*(15), 1–7. https://doi.org/10.1007/s13347-023-00621-y
Gaudet, M. J., Herzfeld, N., Scherz, P., & Wales, J. J. (Eds.). (2024). *Encountering Artificial Intelligence: Ethical and Anthropological Investigations.* AI Research Group for the Centre for Digital Culture of the Dicastery for Culture and Education of the Holy See. Eugene, Oregon: Pickwick Publications.
Gottfredson, L. (1997). Mainstream Science on Intelligence: An Editorial With 52 Signatories, History, and Biography. *Intelligence, 24*(1), 13–23.
Han, B.-C. (2017). *The Swarm.* The MIT Press.
Han, B.-C. (2022). *Non-things.* Polity Press.
Hintikka, J. (1980). Aristotelian Induction. *Revue Internationale de Philosophie, 34*(133–134), 422–439.
Hintikka, J. (1992). The Concept of Induction in the Light of the Interrogative Approach to Inquiry. In J. Earman (Ed.), *Inference, Explanation, and Other Frustrations: Essays in the Philosophy of Science.* University of California Press.
Jaki, S. L. (1969). *Brain, Mind and Computers.* Herder and Herder.
Julia, L. (2019). *L'intelligence Artifiocielle n'existe pas.* First Éditions.
Kissinger, H. A. (2018, June). How the Enlightenment Ends. *The Atlantic.* Retrieved August 3, 2024, from https://www.theatlantic.com/magazine/archive/2018/06/henry-kissinger-ai-could-mean-the-end-of-human-history/559124/.
Kiverstein, J., Farina, M., & Clark, A. (2013). The Extended Mind Thesis. *Oxford Bibliographies.* Retrieved May 10, 2025, from https://www.oxfordbibliographies.com/display/document/obo-9780195396577/obo-9780195396577-0099.xml.
Landgrebe, J., & Smith, B. (2023). *Why Machines Will Never Rule the World.* Routledge.
Larson, E. (2021). *The Myth of Artificial Intelligence. Why Computers Can't Think the Way We Do.* Harvard University Press.
Lear, J. (1988). *Aristotle, the Desire to Understand.* Cambridge University Press.
Legg, S., & Hutter, M. (2007). A Collection of Definitions of Intelligence, IDSIA-07–07 Technical Report. Retrieved February 22, 2024, from https://arxiv.org/abs/0706.3639.
Mitchell, M. (2019). *Artificial Intelligence. A Guide for Thinking Humans.* Farrar, Straus and Giroux.
Nida-Rümelin, J., & Weidenfeld, N. (2022). *Digital Humanism. For a Humane Transformation of Democracy, Economy and Culture in the Digital Age.* Springer.
Niiniluoto, I. (1999). Defending Abduction. *Philosophy of Science, 66.*, Supplement (Proceedings), S43–S451.
Olezza, M. (2025). Analogical Mappings of Facts and Counterfactuals in the Human Mind and Peirce's Abduction: Limitations in LLMs. *Cognitive Systems*

Research, 94, https://www.sciencedirect.com/science/article/abs/pii/S1389041725000889.

Paavola, S. (2005). Peircean Abduction: Instinct or Inference? *Semiotica, 153*(1/4), 131–154.

Pacani, A. (2024). Anthropomorphism in AI: Hype and Fallacy. *AI and Ethics, 4*, 691–698.

Peirce, C. S. (1958). *Collected Papers* (C. Hartshorne, P. Weiss, A. Burks, Eds.). The Belknap Press of Harvard University Press, Cambridge (1931–1958).

Polanyi, M. (1966) [2009]. *The Tacit Dimension*. With a New Foreword by Amartya Sen. Chicago and London: The University of Chicago Press.

Polanyi, M. (1952). The Hypothesis of Cybernetics. *The British Journal for the Philosophy of Science, 2*(8), 312–315.

Polanyi, M. (1958). *The Study of Man*. The University of Chicago Press.

Porebski, A., & Figura, J. (2025). There is no such thing as conscious artificial intelligence. *Humanities & Social Sciences Communications, 12*, 1647. https://doi.org/10.1057/s41599-025-05868-8

Reeve, C. D. C. (1995). *Practices of Reason. Aristotle's Nicomachean Ethics*. Clarendon Press.

Reeve, C. D. C. (2006). Aristotle on the Virtues of Thought. In R. Kraut (Ed.), *The Blackwell Guide to Aristotle's Nicomachean Ethics* (pp. 199–217). Oxford University Press.

Rufiner, H. L., et al. (2025). Non-Human Agents: Exploring Modern AI Through the Lens of Contemporary Ontology and Theology. *Theology and Science, 23*, (3). https://doi.org/10.1080/14746700.2025.2514297

Searle, J. R. (1994). *The Rediscovery of the Mind*. The MIT Press.

Shields, C. (2002). *Order in Multiplicity*. Oxford University Press.

Shields, C. (2014). *Aristotle* (2nd ed.). Routledge.

Sunday Grève, S. (2025). Turing's Philosophy of AI. *Journal for General Philosophy of Science, 56*, 513–529.

Tozer, T. (2020). What Computers Will *Never* Be Able To Do. In R. Skidelsky & N. Craig (Eds.), *Work in the Future. The Automation Revolution* (pp. 99–108). Palgrave Macmillan.

Turing, A. M. (1950). Computing Machinery and Intelligence. *Mind NS, 59*(236), 433–460.

Vallor, S. (2024). *The AI Mirror. How to Reclaim our Humanity in an Age of Machine Thinking*. Oxford University Press.

Vigo, A. (2006). *Estudios aristotélicos*. EUNSA.

Weizenbaum, J. (1976). *Computer Power and the Human Reason*. Freeman and co.

CHAPTER 6

The Impact of AI on Our Identity, Society, Culture, Economy, and Politics

Abstract This chapter assesses the impact of AI developments on human identity and freedom, as well as on society, work, the economy, healthcare, education, and the environment. There are both positive and negative effects that call for specific and urgent actions. Indeed, the chapter concludes that we face a paramount challenge: finding and implementing ways to transform the impact of new technologies into beneficial contributions to the human and social good.

Keywords AI Impact • Identity • Society • Labour • Politics

We are witnessing a watershed change: indeed, we face the introduction of a new way of life with a deep anthropological impact, as a new social order ushers the 'algorithmic society' in. Jack Balkin has described this new cultural and political organization as 'a society organized around social and economic decision-making by algorithms, robots, and AI agents' [1, 1151 nt. 1]. AI has borne and will continue to bear cultural, political, and economic effects, bringing opportunities and challenges that we will need to prepare to manage positively for the good of mankind and society, with respect for human dignity. As Miles Brundage asserts, 'if humanity successfully navigates the technical, ethical and political challenges of

R. F. Crespo, *A Classical Philosophical Approach to Artificial Intelligence*, https://doi.org/10.1007/978-3-032-18462-7_6

developing and diffusing powerful AI technologies, AI may have an enormous and potentially very positive impact on humanity's wellbeing' (2018, p. 13). In effect, AI's capacity to process enormous amounts of data at staggering speed and its ability to draw conclusions from elaborating them would be fantastically helpful for human beings, greatly enlarging our operational and calculating capabilities. Furthermore, for Shannon Vallor (2024, pp. 135ff. and 172), AI can help to overcome economic, geopolitical, and environmental crises. A lot of benefits yielded by the combination of AI and big data are currently evident, already making our daily life easier. However, as Jean Ladrière warns, technical radical changes produce a metamorphosis of the whole setting of life, and every metamorphosis features a destructuration phase (1977, p. 69).

Assessing the anthropological impact of AI requires a definition of the human nature that would be probably affected. In the previous chapter, I have taken into account the Aristotelian notion of human beings as *zôon lógon échon*, animals with language capabilities—i.e., possessing a rational character that includes reason, intelligence, rational will, and emotions guided by reason. In the same chapter of *Politics* I, Aristotle viewed human being as *zôon politikón*, as noted before (I, 2, 1253a 11–18):

> Now, that man is more of apolitical animal than bees or any other gregarious animals is evident. Nature, as we often say, makes nothing in vain, and man is the only animal whom she has endowed with the gift of speech. And whereas mere voice is but an indication of pleasure or pain, and is therefore found in other animals (for their nature attains to the perception of pleasure and pain and the intimation of them to one another, and no further), the power of speech is intended to set forth the expedient and inexpedient, and therefore likewise the just and the unjust. And it is a characteristic of man that he alone has any sense of good and evil, of just and unjust, and the like, and the association of living beings who have this sense makes a family and a state.

So, it is characteristic of human beings to share this sense of good and evil, just and unjust, thus building a society. Humans are relational beings that have common values, joys, pains, and sufferings, which binds them together in a relation that entails a sincere interest in each other, as well as in their ideas and projects. Then, as I will discuss in Chap. 8, another characteristic of human beings is the freedom of their rational will: we have an inner ability to decide how to act.

Finally, every human being has a personal, nontransferable identity, which is deeply rooted and goes far beyond a mere name. While we are all humans, we are all unique, and our individual identity has accidental specific properties. Human identity encompasses the 'radical' or 'basic' *who* a person is and also their subsequent characteristics and actions that complete this *who*. In Crespo (2019), I introduced an explanation of identity—John Finnis penned it (2005)—based on Aristotle's hylomorphism, a general doctrine whereby all substances are composed by matter and form as two co-principles that are inseparable and indistinguishable within any specific individual substance. Aristotle applied this doctrine to human beings, who are rational ensouled bodies—the soul being the form and the body, the matter—that naturally develop the characteristics and abilities connecting the physical and the intentional. Finnis describes four orders of reality for personal identity: the theoretical (better yet, ontological), the logical (gnoseological), the moral, and the technical orders. The first one corresponds to human beings' hylomorphic constitutive condition, which encapsulates their diachronic identity—their identity over time—and their synchronic identity—the difference between every individual in the same species. Form and matter—soul and body—together ensure humans' diachronic and synchronic identity, creating every unique individual human being. The second order refers to the identity of knowers, while the third portrays the identity of practical reasoning beings that shape themselves according to their reasoning. Finally, concerning the fourth order, 'the persona is an artefact, intelligible and assessable as a means of communication, and by all the criteria applicable to the arts' (Finnis, 2005, p. 252). This order may seem reductivist, but it is inseparable from the others.

The last three orders naturally link personal identity to social identity. Finnis (2005, p. 280) quotes Germain Grisez (1975, pp. 349 and 351):

> A person is in all four of the orders, and he embraces all of them in himself. In the person the four orders are distinct, irreducible, yet normally inseparable. [...] The four aspects of the person all involve and in a way include one another, as the four orders always do... The person is the self who *unifies* these four distinct and irreducible but normally inseparable aspects. The self [the soul] is a unifying principle.

These orders are distinct but united in the same person, while the hylomorphic identity—the first order—provides the foundation on which the other orders are built. In turn, AI bears an impact on all these human characteristics: human beings' personal identity, social nature, rationality—including emotions—and freedom.

For starters, concerning anthropological identity, AI shapes a person's individual identity poorly, in a way that strays far from the rich notion described above. As Byung-Chul Han (2022b, p. 14) observes, 'Identity is today produced through information. We *produce ourselves* on social media.' We create profiles for ourselves in these media, and our profiles are influenced by these very media, often not necessarily matching our real identities. Our photographs on these media do not represent our everyday life, while the massive information about ourselves that we almost unconsciously provide is used to suggest products, experiences, and content that are gradually moulding our personality. AI also shapes people's profile, their 'digital identity,' which may determine their opportunities for employment, credit, healthcare, etc.[1] Many years ago, AI scientist Joseph Weizenbaum (1972, 1976, p. 26) also warned about the dehumanizing tendency to think of human beings as simple computer-like entities, pointing to the dehumanizing ethos that fuelled AI's development (1976, pp. 140 and 266).

In his several books, Han also stresses the detachment from reality produced by the current information overload —'the unfiltered mass of information is dulling our senses' (2017, p. 59)—as our attention scatters around. The eloquent title of his book *Non-things* (2022b) foreshadows his warning about 'our intoxication by communication and information that makes things disappear' (2022b, p. viii). 'The digital order *de-reifies* the world by *informatizing* it [...] Information, rather than things, determines the lifeworld' (2022b, p. 1). Actual, existent things are messages and images, as we witness a crisis of facts, a decay of truth (2022a, pp. 44, 50–51), and live in a state of 'truthiness'—the quality of being true because we feel or want it to be true, independently of facts (2022a, p. 47). Digital communication replaces thinking, which requires the calm and quiet that are being destroyed by over-information. Éric Sadin (2015, pp. 117 and 124) also refers to the reduction of reality implied by the universal quantification operated by AI.

This distance from reality impacts on our relationship with others. AI could and in fact does contribute to communication, detecting those in

[1] Gloria Andrada (2025), for example, analyses the cognitive impact on people's identity of AI beautifying filters.

need and promoting ways to help them. However, Han argues that the nature of the links established by AI goes in a different direction. In *In the Swarm* (2017), this author describes the characteristics of our digital society, which he calls 'the digital swarm,' highlighting its lack of unity and soul. There is no 'we,' no harmony in this swarm that lacks the interiority and the inner life of a group of human beings. The disruptive impact of technique produces people's uprooting (Ladrière, 1977, p. 83), and technology, Ladrière states, seems unable to provide the unifying principle of culture (1977, p. 99). Every individual has their own private identity and is isolated from others—e.g., people participate in several chats that are inherently unstable (2017, pp. 9–12, 22–24). In *Non-things* Han also insists on the disappearance of the other as a gaze and voice, the absence of empathy (2022b, pp. 20–21, 53–55, 73), stressing that AI does away with both caring (2022b, p. 4) and happiness (2022b, p. 94). While Han's opinions should probably be taken with a grain of salt, because perhaps not the entire population suffers from the effects he describes, they do point to a real tendency. Vallor (2024, pp. 108–110) also remarks that the AI system is threatening the 'space of moral reasons,' borrowing from John McDowell's notion to refer to the realm where we formulate and share moral values. She writes, 'The space of moral reasons is shrinking, both personally and in the public life, as a result of our growing reliance on increasingly opaque and automated machine decisions' (2024, p. 110).

The anthropomorphizing language used to refer to AI applications' relations with humans tends to prove confusing, but, when analysed in deep, these relations are not genuinely human, as AI does not express real empathy for or interest in people. Its gestures are merely facial expressions or phrases that do not build up or grow into actual personal connections. Besides, as Adriana Placani argues, 'anthropomorphism distorts moral judgements through its fallacious character' (2024, p. 691).

With AI's deterministic nature, individuals in the digital swarm might as well live in the context of Max Weber's theory of disenchantment, feeling somewhat beguiled by AI's 'magical' capabilities, in what Alexander Campolo and Kate Crawford call an 'enchanted determinism' (2020, p. 1).

Ladrière also notes that AI arises at a time when everyone is everywhere and nowhere at the same time, in an era where nothing has significance—a period of disenchantment (1977, p. 84). Indeed, AI plays an active role in the modern (terrible and depressing) project of rationalization of individual lives and societies by producing automatic reactions to the indications stemming from the work of algorithms with big data. Thus, data become performative (cf. Sadin, pp. 62ff.), and AI threatens freedom. Ladrière warned that, while expanding men's power to act, the 'automata' increasingly shaped day-to-day life (1977, p. 79).

'Surveillance capitalism,' an expression coined by Shoshana Zuboff (2015, 2019), exploits freedom (2022b, p. 25), although it is more of a kind of self-exploitation. We are continuously and freely providing information about our opinions, preferences, and desires, and algorithms use that data to offer suggestions that—slowly but surely—chip away at our autonomy. Yet, we still feel as if we were free, but this freedom is an illusion. Han states, 'today, thinking ourselves free, we are in fact captives in a *digital* cave' (2022a, p. 58).

Thanks to the ubiquity of information (Han, 2017, p. 34), we also have the compulsion to work always and everywhere. The prevailing language revolves around performance and efficiency, while effectiveness has replaced truth (Han, 2022b, p. 6). Rob Kitchin (2017, p. 20) states that 'Algorithms are used to seduce, coerce, discipline, regulate and control: to guide and reshape how people, animals and objects interact with and pass through various systems.' Data, according to Balkin (2018, p. 1157), 'powers the engines of the Algorithmic Society.' As a result, data provides control to the group of people, governments, corporations, and companies that use them. Indeed, Zuboff's (2015, 2019) 'surveillance capitalism' refers to how big corporations are using algorithms and big data to predict users' behaviours in order to offer products, trying to mould human behavioural patterns to secure financial results. According to her, today's capitalism creates a new economic order that uses human experience as free raw material to extract information so as to predict and promote sales. Knowledge and freedom are controlled by big tech corporations, while a significant share of power is held by a few large AI companies.[2] Sadin (2015, pp. 199ff.) calls it the 'techno-power' and shows its huge penetration in every aspect of life and society to raise those corporations' profits. This power concentration poses a risk of data manipulation for personal, corporate, or political interests. The alliance among technology, economy, and efficiency does not necessarily favour the human good.

In another book on the political consequences of AI, *Infocracy* (2022a), Han speaks of a 'tribalization' of society, with tribes that do not communicate with each other and lead to opinion and identity dictatorships (2022a, p. 33). We seem to have become deaf to the voices of others

[2] Nick Srnicek (2020) explains the monopolization tendency of big platforms.

(2022a, p. 30), while data surveillance also attempts to gain control of our political decisions. In this book, Han denounces 'the information regime', 'a form of domination in which information and its processing by algorithms and artificial intelligence have a decisive influence on social, economic and political processes' (2022a, p. 1). The influence exerted on our opinions and choices often lurks below our consciousness (2022a, p. 10), as the pressure to make quick decisions and the overwhelming amount of information often drive us to react (rather than to act) automatically, following AI's suggestions. 'Dataists' view AI as a better listener than humans (2022a, p. 35), even daring to propose the replacement of politics by deciding processes led by AI. As Brundage (2018, p. 17) states, 'Such a technology could be abused to create a stable authoritarian state of unprecedented endurance and global scale, relying on automation of surveillance, coercion and the crushing of dissent.'

AI may also bear a significant impact on the future of work, with pessimistic views claiming that AI will replace millions of jobs, while optimistic observers contend that new alternatives will emerge. Some of the latter even go as far as to hint at a new and overarching notion of work. The ideas of the 'positive psychology' school of thought as well as the distinctions among 'employment,' 'career,' and 'vocation' as forms of work are some keys to this broader notion of labour.

Positive psychology arises as an expansion of the focus of psychology on pathologies towards the study of positive experiences and traits that lead to well-being. It maintains that happiness is associated with the experience of giving one's best. Mihaly Csikszentmihalyi and Martin Seligman have made a great contribution by starting this psychological current. In addition, American sociologist Robert Bellah led a group of researchers in a study of American people's character (1985). For these researchers, attitudes towards work indicate how private and public lives are related, and they identified three possible attitudes. For those who view work as a 'job,' it is a way to make a living. In turn, people who think of work as a 'career' seek to stand out and succeed in their field. Finally, those who work to pursue a 'vocation' feel their work is a mission that identifies them and gives meaning to their life.

The unprecedented conditions of the new era that is coming, some optimistic thinkers argue, allow for the combination of positive psychology's ideal of happiness and work as a calling. Tasks that were mostly carried out as jobs are going to be replaced by AI, and people are going to concentrate on endeavours that enable them to achieve this combination of positive happiness and vocational pursuits. This type of work is linked to creativity and contributions to society (see Carl B. Frey, & Michael A. Osborne, 2017, p. 269). The so-called soft abilities— empathy, communication, sensibility, leadership—will take the spotlight, and work will make us more human, helping us to display our identity. We need an attractive, challenging—but not stressful—job, and this becomes a very real possibility in this new era, according to these thinkers. The key lies in choosing what tasks we want AI to take on for us and what will remain part of our roles. In other words, we should be able to view AI as a complement, not a replacement.

However, pessimistic observers believe the speed of current changes prevents the labour force's adaptation to new forms of working, leaving people to carry out repetitive, more basic tasks. Amidst such rapid shifts, education and training programs are likely to consistently lag behind AI developments, while many jobs that are enriched by human beings' creativity and dedication will probably be reduced to standard processes with efficiency as the ruling criterion. The people who have influenced the course of human history more decisively were not technicians but intellectuals, as noted by Eric Larson (2021, pp. 237–281), who underscores the need of science and thinkers for the future of AI. The negative impact might also hit individuals working in the AI sector. Jenna Burrell and Marion Fourcade (2021) point to the rise of a 'coding elite' of software developers, investors, professors, etc., as well as to the emergence of a 'cybertariat' consisting of workers tasked with the unskilled job of correcting mistakes, adapting programs, etc.[3]

Concerning AI's impact on the economy, Daron Acemoglu and Pascual Restrepo (2019) argue that automation generates a displacement effect that might be balanced or not by the consequences of increased productivity on the demand for labour from non-automated tasks. In effect, as Daniel Susskind (2020) points out, AI will have a lesser impact on non-routine tasks than on routine jobs. In a more recent book (2023), Acemoglu and Simon Johnson highlight the concentration of power of AI corporations that look for benefit maximization and must be staved off by the State and labour unions, but they also note that AI tools cannot execute tasks involving 'social interaction, adaptation, flexibility and communication' (2023, p. 315) as humans do. They posit that we have to seek labour force complementation rather than replacement, looking for

[3] See Crawford (2021), Chapter 2.

more prosperity for workers. They also suggest a set of specific actions to control AI development (2023, Chapter 11). Acemoglu's Nobel Lecture summarizes his work with Simon Johnson and James Robinson on AI's amplifying effect over automation, stressing the importance of choosing right institutions to guide its progress (2025a, pp. 1740–1742; see also 2025b). Anton Korinek and Joseph Stiglitz (2025) also recommend some specific actions to care for human labour, whereas Irmgard Nübler (2020, p. 195) proposes 'a wide range of policy issues in relevant fields: learning of individuals and societies; guiding science, technology and innovation for job creation; and enhancing labour market dynamics.'[4]

A universal basic income—even if it were feasible—is not the solution, because work brings purpose to our life, making it necessary for a fulfilled life—something that really matters and cannot be cast away. As Julian Nida-Rümelin and Nathalie Weidenfeld put it, 'there is more to a good life than being fed' (2022, p. 106). Capital will be increasingly relevant given the resources necessary to develop AI, and the effects on work are likely to be mixed: some will succeed in transforming themselves, and some will fail to do it. Marina Tavares stresses that 'the challenge for policymakers is to create the conditions that will allow AI to enhance human potential' (2025, p. 27). Based on several research studies, she concludes that the impact will not be evenly distributed, which will call for actions to make sure its benefits trickle down to everyone. Robert Skidelsky also worries about the distributive effects of AI development and thinks that new policies will be needed, and so do I, if we want human labour to survive. Yet, Skidelsky points to a deeper problem, remarking, 'This seems to me to be as much as policy can do. None of this, though, addresses the question raised by Patočka: how humans can be enabled to 'feel at home' in world governed by an inhuman and, if not intentionally, inhumane, logic' (2020, p. 21).

A field that is particularly affected by AI is healthcare. Its professionals have a profound vocation that translates into the care and respect for

[4] The World Economic Forum Future of Work Report 2025 (https://www.weforum.org/publications/the-future-of-jobs-report-2025/) estimates that, by 2030, there will be significant job creation (170 million) and displacement (92 million), with growth in roles related to AI, green technologies, and healthcare. A shift in workforce demographics and increased focus on diversity, equity, and inclusion are also expected. By 2030, employers plan to prioritize employee well-being, talent retention, and investing in AI-related skills. These 'predictions' imply institutional policymaking to prepare workers to adapt successfully to these changes.

human life. AI can and in fact does help to diagnose diseases and suggesting treatments. However, final decisions should be made by healthcare professionals, whose personal relationships with patients cannot be replaced. Another risk that must be taken into account is the invasion of people's privacy by large healthcare providers, whose guidelines and standards are geared towards maximizing financial results (see Sadin, 2015, p. 91).

AI can also contribute to education, and it is already doing that. There are very good experiences—such as Khan Academy's use of AI—that must be supported (see Salman Khan, 2024), but human professors cannot be replaced. Education is not a mere transfer of knowledge: its aim is human beings' overall development and growth. The relationship between teachers and students cannot be replaced by AI applications, which should only contribute to and support it. Neither should students' intellectual work be carried out by AI, as this would lead to their increasing impoverishment. Education should help people to profit from AI wisely, dealing, more than ever, with deep philosophical questions and separating wisdom from knowledge.

Other widely pointed out impacts for people and societies notably include misinformation, automatic propaganda, deepfakes, fake news, micro targeting, biased judgments, privacy and reputational risks. The mediation of AI in communication decreases epistemic trust, perceptions of authenticity, responsibility, and competence (cf. Sahebi & Formosa, 2025), as information alone is unstable and does not build trust.

The impact on the environment should also be considered, given the extraordinary demand of mineral resources, water, and energy that the digital devices, storage of data and functioning of AI applications, especially generative AI, imply (cf. Crawford, 2021, Chapter 1 and Luc Julia, 2020, p. 122). Some of these problems might be mitigated with control and technological advances, as we shall discuss in the next chapter when we tackle AI control, laws, and ethics.

In sum, we face a paramount challenge: finding ways to transform the impact of new technologies in positive contributions to the human and social good.

References

Acemoglu, D. (2025a). Nobel Lecture: Institutions, Technology, and Prosperity. *American Economic Review, 115*(6), 1709–1748.

Acemoglu, D. (2025b). The Simple Macroeconomics of AI. *Economic Policy, 40*(121), 13–58. https://doi.org/10.1093/epolic/eiae042

Acemoglu, D., & Johnson, S. (2023). *Power and Progress: One Thousand-Year Struggle Over Technology and Prosperity*. Public Affairs.

Acemoglu, D., & Restrepo, P. (2019). Automation and New Tasks: How Technology Displaces and Reinstates Labor. *Journal of Economic Perspectives, 33*(2), 3–30.

Andrada, G. (2025). Beauty Filters in Self-Perception: The Distorted Mirror Gazing Hypothesis. *Topoi, 44*, 367–378. https://doi.org/10.1007/s11245-024-10149-1

Aristotle. *Politics* (B. Jowett, Trans.). Retrieved September 3, 2024, from https://classics.mit.edu/Aristotle/politics.1.one.html.

Balkin, J. (2018). Free Speech in the Algorithmic Society: Big Data, Private Governance, and New School Speech Regulation. *UC Davis Law Review, 51*(3), 1149–1210.

Bellah, R. N., Madsen, R., Sullivan, W. M., Swidler, A., & Tipton, S. M. (1985). *Habits of the Heart*. Harper & Row.

Brundage, M. (2018). Scaling Up Humanity: The Case for Conditional Optimism about Artificial Intelligence. In European Parliamentary Research Service. *Should We Fear Artificial Intelligence?* Retrieved 3 September 2024, from https://www.europarl.europa.eu/RegData/etudes/IDAN/2018/614547/EPRS_IDA(2018)614547_EN.pdf.

Burrell, J., & Fourcade, M. (2021). The Society of Algorithms. *Annual Review of Sociology, 47*, 213–237.

Campolo, A., & Crawford, K. (2020). Enchanted Determinism: Power without Responsibility in Artificial Intelligence. *Engaging Science, Technology, and Society, 6*. https://doi.org/10.17351/ests2020.277.

Crawford, K. (2021). *Atlas of AI*. Yale University Press.

Crespo, R. F. (2019). Identity: Individual and Social, and a Theory of a 'Polymorphic' but Unique Identity. In W. Dolfsma, W. Hands, & R. McMaster (Eds.), *History, Methodology and Social Economics* (pp. 51–66). Routledge.

Finnis, J. M. (2005). The Thing I am': Persona Identity in Aquinas and Shakespeare. *Social Philosophy and Policy, 22*(2), 250–282.

Frey, C. B., & Osborne, M. A. (2017). The Future of Employment: How Susceptible Are Jobs to Computerization? *Technological Forecasting & Social Change, 114*, 254–280.

Grisez, G. (1975) Beyond the New Theism: A Philosophy of Religion (Notre Dame: University of Notre Dame Press).

Han, B.-C. (2017). *The Swarm*. The MIT Press.

Han, B.-C. (2022a). *Infocracy*. Polity Press.

Han, B.-C. (2022b). *Non-things*. Polity Press.

Julia, L. (2020). *L'intelligence artificielle n'existe pas*. First Éditions.

Khan, S. (2024). *Brave New Words.* Viking.
Kitchin, R. (2017). Thinking Critically About and Researching Algorithms. *Information, Communication & Society, 20*(1), 14–29. https://doi.org/10.1080/1369118X.2016.1154087
Korinek, A., & Stiglitz, J. (2025). Steering Technological Progress. INET Working Paper No. 232. Retrieved May 30, 2025, from https://www.ineteconomics.org/uploads/papers/WP-232-SteeringProgress-May-2025.pdf.
Ladrière, J. (1977). *The Challenge Presented to Cultures by Science and Technology.* UNESCO.
Larson, E. (2021). *The Myth of Artificial Intelligence. Why Computers Can't Think the Way We Do.* The Belknap Press of Harvard University Press.
Nida-Rümelin, J., & Weidenfeld, N. (2022). *Digital Humanism. For a Humane Transformation of Democracy, Economy and Culture in the Digital Age.* Springer.
Nübler, I. (2020). Shaping the Work of the Future: Policy Implications. In R. Skidelsky & N. Craig (Eds.), *Work in the Future. The Automation Revolution* (pp. 189–201). Palgrave Macmillan.
Pacani, A. (2024). Anthropomorphism in AI: Hype and Fallacy. *AI and Ethics, 4,* 691–698.
Sadin, É. (2015). *La Vie Algorithmique. Critique de la Raison Numérique.* Éditions L'Échappée.
Sahebi, S., & Formosa, P. (2025). The AI-Mediated Communication Dilemma: Epistemic Trust, Social Media, and the Challenge of Generative Artificial Intelligence. *Synthese, 205,* 128. Retrieved April 2, 2025, from https://philarchive.org/archive/SAHTAC
Skidelsky, R. (2020). The Future of Work. In R. Skidelsky & N. Craig (Eds.), *Work in the Future. The Automation Revolution* (pp. 9–22). Palgrave Macmillan.
Srnicek, N. (2020). Two Myths About the Future of the Economy. In R. Skidelsky & N. Craig (Eds.), *Work in the Future. The Automation Revolution* (pp. 133–142). Palgrave Macmillan.
Susskind, D. (2020). Work in the Digital economy. In R. Skidelsky & N. Craig (Eds.), *Work in the Future. The Automation Revolution* (pp. 125–132). Palgrave Macmillan.
Tavares, M. (2025). A Place for Human Talent in the AI Age. *Finance and Development, IMF, March, 2025,* 27–29.
Vallor, S. (2024). *The AI Mirror. How to Reclaim our Humanity in an Age of Machine Thinking.* Oxford University Press.
Weizenbaum, J. (1972). On the Impact of the Computer on Society. *Science, New Series, 176*(4035), 609–614.
Weizenbaum, J. (1976). *Computer Power and Human Reason.* Freeman and Co.

Zuboff, S. (2015). Big Other: Surveillance Capitalism and the Prospects of an Information Civilization. *Journal of Information Technology, 30*, 75–89. https://doi.org/10.1057/jit.2015.5

Zuboff, S. (2019). *The Age of Surveillance Capitalism: The Fight for a Human Future at the New Frontier of Power*. Public Affairs.

CHAPTER 7

Remedies

Abstract The conclusion drawn in the previous chapter clearly calls for actions to promote the good use of AI in favour of humankind and society's common good. This chapter analyses the role of international organizations and their declarations, the role of the State and its regulations, the role of AI corporations themselves and their codes of ethics, as well as the means and hurdles for AI accountability regarding its effects. Like its predecessors, the chapter starts by introducing Aristotle's thoughts on the role played by the State (the *polis*) to promote human beings' and society's welfare.

Keywords The Role of the State • AI Declarations • AI Regulations • AI Ethical Codes • AI Accountability

The conclusion of the previous chapter clearly calls for actions to promote the good use of AI in favour of humankind and society's common good. Now the time has come to analyse the role of international organizations with their declarations, the role of the State with its regulations, the role of AI corporations themselves with their codes of ethics, and the means and hurdles for AI accountability regarding its effects. This chapter, like its predecessors, will start by introducing Aristotle's thoughts on the role of the State (the *polis*) concerning human beings' and society's good. In

R. F. Crespo, *A Classical Philosophical Approach to Artificial Intelligence*, https://doi.org/10.1007/978-3-032-18462-7_7

turn, the next chapter will discuss whether developing 'moral machines' would provide another possible remedy for the AI system's undesired effects. Then, Chap. 9 will stress the need for AI developers' and users' personal virtue and commitment to guiding it towards the promotion of human and social good.

Aristotle on the Role of the State

For Aristotle, the good of man is the same as the good of the *polis*, and this good is to live the Good Life that leads to happiness (e.g., 'the best way of life, for individuals severally as well as for states collectively, is the life of goodness,' *Politics* VII, 1, 1323b 40–41; cf. *Politics* VII, 2, 1323a 5–8—the felicity of the state is the same of the felicity of the individual; *NE* I, 2, 1094b 7–8). 'The polis,' Aristotle wrote, 'is an association [*koinonía*] of freemen' (*Politics* III, 6, 1279a 16). And what is the end of this association? He replied:

> It is clear, therefore, that a polis is not an association for residence on a common site, or for the sake of preventing mutual injustice and easing exchange. These are indeed conditions which must be present before a polis can exist; but the existence of all these conditions is not enough, in itself, to constitute a polis. What constitutes a polis is an association of households and clans in a good life [*eû zên*], for the sake of attaining a perfect and self-sufficing existence. (*Politics* III, 9, 1280b 29–35)

Thus, the task of political communities and society's officials is to drive and facilitate the good actions that enable all citizens to attain this life of true happiness and goodness. Three quotations in this regard prove enlightening:

> Political science spends most of its pains on making the citizens to be of a certain character, viz. good and capable of noble acts (*NE* I, 9, 1099b 30–31). To have a character good and capable of noble acts is to be virtuous.
>
> There is one thing clear about the best constitution: it must be a political organization which will enable all sorts of men [e.g., both 'contemplative' and 'practical' individuals] to be at their best and live happily [*árista prắttoi kaì zóe makaríos*] (*Politics* VII, 2, 1324a 23–25; quoted also by Martha Nussbaum, 1987, p. 2).
>
> The true end which good law-givers should keep in view, for any state or stock or society with which they may be concerned, is the enjoyment of

partnership in a good life and the felicity [*zoês agathês* ... *kaì* ... *eudaimonías*] thereby attainable. (*Politics* VII, 2, 1325a 7–10; quoted also by Nussbaum, 1987, p. 3)

These last two statements lead Nussbaum to uphold that 'the task of political arrangement is both broad and deep' (1987, p. 6, 1990, p. 209).

I have explained (Crespo, 2007, p. 376) the classical Aristotelian distinction among a) ends that can be considered only as means, pursued solely for the sake of something else (first-order or instrumental ends); b) ends that are desirable in themselves and for the sake of the final end (second-order ends); and c) ends which are only desirable in themselves (third-order or final ends: usually known as 'happiness'). I have also provided the following example: we study for an exam (i.e., a means to an instrumental end) in order to graduate (a second-order end) and be happy (a final end), according to our life plan (designed by practical reason). Practical rationality harmonizes the complex set of second-order ends in order to follow a plan that will lead us to happiness. What are, for Aristotle, the second-order ends that contribute to a happy life? In *Nicomachean Ethics,* he mentions honour, wisdom, and pleasure (I, 6, 1096b), later adding reason (*noûn*) and every virtue (I, 7, 1097b 2).

A virtuous person, a person who rightly exercises their practical reason knows how to combine the elements at hand—even if something is lacking—in order to be happy, meaning that both practical reason and virtue hold the keys to happiness. Hence, in *Nicomachean Ethics,* Aristotle noted, 'legislators make the citizens good by forming habits in them, and this is the wish of every legislator, and those who do not effect it miss their mark, and it is in this that a good constitution differs from a bad one' (II, 1, 1103b 3–6).

For Aristotle, law-makers foster citizen's virtues via education and laws: virtues, laws, and education build a virtuous circle that makes people happy, ensuring their self-realization and the stability of the political regime. Laws are obeyed by virtuous people, and people are not virtuous if they have not been educated since youth, which takes us right back to laws, as they must support education. Aristotle stated, 'The law bids us practice every virtue and forbids us to practice every vice. And the things that tend to produce virtue taken as a whole are those of the acts prescribed by the law which have been prescribed with a view to education for the common good' (*NE* V, 2, 1130b 23–27). He expounded:

> it is difficult to get from youth up a right training for virtue if one has not been brought up under right laws; for to live temperately and hardily is not pleasant to most people, especially when they are young. For this reason their nurture and occupations should be fixed by law; for they will not be painful when they have become customary. But it is surely not enough that when they are young they should get the right nurture and attention: since they must, even when they are grown up, practise and be habituated to them, we shall need laws for this as well, and generally speaking to cover the whole of life; for most people obey necessity rather than argument, and punishments rather than the sense of what is noble. This is why some think that legislators ought to stimulate men to virtue and urge them forward by the motive of the noble, on the assumption that those who have been well advanced by the formation of habits will attend to such influences; and that punishments and penalties should be imposed on those who disobey and are of inferior nature, while the incurably bad should be completely banished. (...) the law *has* compulsive power, while it is at the same time a rule proceeding from a sort of practical wisdom and reason. (*NE* X, 9, 1179b 20–1180a. 22)

Political institutions are designed to achieve people's happiness. 'The end and purpose of a polis is the good life, and the institutions of social life are means to that end' (*Politics* III, 9 1280b 39–40). Aristotle elaborated extensively on the different ways to elect assemblies, magistrates, courts, explaining people's engagement in the entire process (*Politics* IV, 14 and ff.).

Education proves instrumental to preserving these institutions: 'The greatest, however, of all the means we have mentioned for ensuring the stability of constitutions—but one that nowadays is generally neglected—is the education of citizens in the spirit of their constitution. There is no profit of the best of laws, even when they are sanctioned by general civic consent, if the citizens themselves have not been attuned, by the force of habit and the influence of teaching, to the right constitutional temper' (*Politics* V, 9, 1310a 12–18). Friendship and consensus (concord—*omónoia*) also hold cities together (*NE* VIII, 1, 1155a 22–26; IX, 6, 1167b 2).

Summing up, law and education foster the development of virtues, while living in the pursuit of virtues produces happiness, which is the purpose of a political community. Thus, from an Aristotelian perspective, regulations and codes of ethics are very useful to encourage the people involved in AI development to behave morally, but they are regarded with distrust and as threatening in the AI world.

Declarations, Regulations, and Codes of Ethics

Several authors dwell on the changes and new legal instruments brought about by Big Data, recommending changes to dominant legal frameworks and practices as well as new accountability processes. Hamid Ekbia et al state, 'the complexity and ubiquity of Big Data requires a concerted effort between academics and policy makers, in rich conversation with the public' (2015, p. 26). We can all agree that algorithms must be transparent, comprehensible, predictable, and controllable.

The European Parliamentary Research Service (EPRS, 2018) has elaborated several proposals to manage AI effectively so as to maximize its benefits and avoid the risks it may pose. This document was written by a team consisting of Peter Bentley, Miles Brundage, Olle Hägström, and Thomas Metzinger. Bentley notes, "we should be focusing on new safety regulation and certification for each specific safety-critical application of AI" (2018, p. 19). Brundage believes that solving the control problem is a critical prerequisite for AI systems to have positive impacts on society (2018, p. 14). Metzinger holds that political institutions must produce and implement a minimal but sufficient set of ethical and legal constraints for AI's beneficial use and future development. They should also create a rational, evidence-based critical discussion process to continuously update, improve, and revise this initial set of rules and guidelines. (2018, p. 27)

By early 2025, "Around the world, at least 69 countries have proposed over 1000 AI-related policy initiatives and legal frameworks to address public concerns around AI safety and governance" (Nick Sherman, https://www.mindfoundry.ai/blog/ai-regulations-around-the-world). We can isolate three action paths: first, a group of recommendations, declarations, and initiatives; a second set of laws and regulations, and, finally, a code of ethics. The first group stems from international organizations or meetings, while the second group comes from countries that have enacted specific laws, and the third has been drafted by private corporations working on AI.

Declarations

Let us take a look at the most recent initiatives as of 2025. Their shared values and principles revolve around the defence of human rights and fundamental freedoms—such as privacy, sustainable development, promotion of diversity, and inclusiveness—to ensure the existence of peaceful, fair, and democratic societies. Their common concerns zero in on the likely use of AI for social scoring and mass surveillance, its potential to enable discrimination, its negative environmental impact, and its lack of transparency and explainability.

In April 2019, the first Vienna Workshop on Digital Humanism gathered over 100 members of academia, public institutions, civil society organizations, and business companies. The workshop came to an end with its 'Vienna Manifesto on Digital Humanism' (www.dighum.org/dighum-manifesto/). Both this workshop and the Manifesto it drafted prove interesting because they marked the first step taken by the Digital Humanism Initiative (https://caiml.org/dighum/), which has held more workshops, conferences, lectures, and summer schools, as well as publishing open letters, declarations, and books.[1] Introduced 'as an approach that describes, analyses, and, most importantly, influences the complex interplay of technology and humankind, for a better society and life, fully respecting universal human rights' (Hannes Werthner, 2024, p. 123), this initiative draws its inspiration from the Enlightenment and humanist ideas. Julian Nida-Rümelin and Klaus Staudacher describe ´humanism´ in the following terms: 'At the heart of humanist philosophy and practice is the idea of human authorship. Human beings are authors of their lives; as such, they bear responsibility and are free. Freedom and responsibility are two mutually dependent aspects of human authorship. Authorship, in turn, is linked to the ability to reason' (2024, p. 18).[2]

Another very relevant document is UNESCO's 'Recommendation on the Ethics of Artificial Intelligence' (Recommendation on the Ethics of Artificial Intelligence—UNESCO Biblioteca Digital), which was adopted by the organization's 193 Member States in November 2021. Intended to

[1] Julian Nida-Rümelin and Nathalie Weidenfeld (2022), Hannes Werthner, Erich Prem, Edward A. Lee and Carlo Ghezzi (eds.) 2022, and Hannes Werthner, Carlo Ghezzi, Jeff Kramer, Julian Nida-Rümelin, Bashar Nuseibeh, Erich Prem and Allison Stanger (eds.) 2024.

[2] See Julian Nida-Rümelin and Dorothea Winter (2024) and Julian Nida-Rümelin and Klaus Staudacher (2024) for their interpretation of the meaning of humanism and Enlightenment.

provide a basis to make AI systems work for the good of humanity and to prevent harm, this document proposes policy actions to promote trustworthiness in all the stages of the AI system lifecycle. It calls on member states to roll out effective measures and policy frameworks, making sure that all stakeholders—including private sector companies, academic and research institutions and civil society organizations—adhere to them and encouraging them to develop ethical impact assessment, due diligence tools etc.

Another influential meeting, The AI Safety Summit, held in Bletchley (UK) on 1–2 November 2023, produced 'The Bletchley Declaration' signed by 30 countries (https://www.gov.uk/government/publications/ai-safety-summit-2023-the-bletchley-declaration/the-bletchley-declaration-by-countries-attending-the-ai-safety-summit-1-2-november-2023). This document states that 'AI should be designed, developed, deployed, and used in a manner that is safe, in such a way as to be human-centric, trustworthy and responsible.' Stressing the safety risks that may arise from the 'frontier' of AI, it warns about the possible intentional misuse of AI or its unintended control issues. It also emphasizes the importance of inclusive AI, which should try to bridge the 'digital divide' between and within countries, and highlights the need for responsibility and accountability to ensure the safety of AI throughout its lifecycle.

The G7 meeting's AI Safety Summit in Seoul on 21–22 May 2024 also issued a declaration focusing on AI safety, innovation, and inclusivity. It states, 'We support existing and ongoing efforts of the participants to this Declaration to create or expand AI safety institutes, research programmes and/or other relevant institutions including supervisory bodies, and we strive to promote cooperation on safety research and to share best practices by nurturing networks between these organizations.'[3]

The Global Partnership on Artificial Intelligence (GPAI, https://gpai.ai/) is an international initiative to guide the responsible development and use of AI in a manner that respects human rights and the shared democratic values of its participants. The partnership was first proposed by Canada and France at the 2018 44thG7 summit, and it was officially launched in June 2020. GPAI is hosted by the Organisation for Economic Co-operation and Development (OECD) and has 29 member countries.

[3] https://www.gov.uk/government/publications/seoul-declaration-for-safe-innovative-and-inclusive-ai-ai-seoul-summit-2024/seoul-declaration-for-safe-innovative-and-inclusive-ai-by-participants-attending-the-leaders-session-ai-seoul-summit-21-may-2024.

On *4 December 2024 in Belgrade, Serbia,* the 2024 GPAI Summit took place, hosted by the Government of the Republic of Serbia and supported by the OECD. The Summit gathered GPAI countries, global AI experts, international organizations, industry leaders, and academia to discuss global AI-related challenges and solutions, intending to steer the responsible and human-centric development and use of this technology (Global Partnership on Artificial Intelligence, OECD, 'Belgrade Ministerial Declaration 4 December 2024,' https://wp.oecd.ai/app/uploads/2024/12/GPAI-Belgrade-Declaration-final-3.pdf).

In January 2025, 100 AI experts from 33 countries and intergovernmental organizations prepared a 297-page-long 'International AI Safety Report' (https://www.gov.uk/government/publications/international-ai-safety-report-2025), delving into three types of AI-related risks. First, the report looks at the risks from malicious use, including harm to individuals through fake content, manipulation of public opinion, cyber offence, as well as biological and chemical attacks. Second, it explores malfunctioning risks, such as biases and loss of control. Next, it tackles systemic risks, labour market risks, global AI R&D divide, market concentration and single points of failure, risks to the environment, risks to privacy, and risks of copyright infringement, before, finally reviewing current risk management systems.

The recent AI Action Summit, held in Paris on 11–12 February 2025, issued a 'Statement on Inclusive and Sustainable Artificial Intelligence for People and the Planet' (https://www.elysee.fr/en/emmanuel-macron/2025/02/11/statement-on-inclusive-and-sustainable-artificial-intelligence-for-people-and-the-planet) that was signed by 64 countries. It states, 'We have affirmed the following main priorities:

- Promoting AI accessibility to reduce digital divides;
- Ensuring AI is open, inclusive, transparent, ethical, safe, secure, and trustworthy, taking into account international frameworks for all;
- Making innovation in AI thrive by enabling conditions for its development and avoiding market concentration driving industrial recovery and development;
- Encouraging AI deployment that positively shapes the future of work and labour markets and delivers opportunity for sustainable growth;

- Making AI sustainable for people and the planet, [and]
- Reinforcing international cooperation to promote coordination in international governance.'

Finally, the United Nations' Global Digital Compact (https://www.un.org/global-digital-compact/en), 'Negotiated by 193 Member States and informed by global consultations, (...) commits governments to upholding international law and human rights online and to taking concrete steps to make the digital space safe and secure.' The Compact recognizes the contributions of the private sector, technical communities, researchers and civil society to digital cooperation, while it calls on all stakeholders to engage in realizing an open, safe, and secure digital future for all.

All these declarations, that will surely be followed by other similar declarations in the future, help to raise awareness on AI's relevance and risks, as well as the need for specific regulations. However, these statements must be embedded in countries' legislations to prove effective. With the undeniable economic and political power of AI control, these declarations can become only wishful thinking if they do not translate into concrete laws and sanctions.

Regulations

Several countries have passed specific laws to monitor and control AI developments. These regulations have adopted sectorial approaches (like the United States) or more general outlooks (for example, Canada and China). AI-related legislation has to strike a balance between promoting innovation and upholding ethical standards and societal interests. Some regulatory issues derive from the speed of AI advancements, the autonomy of some AI systems, and AI's increasing opacity (Simon Chesterman, 2021, pp. 6–7). In addition, AI development goes beyond national boundaries and requires international actions.

The EU introduced the world's first comprehensive AI legislation, designed to regulate the use of AI systems across EU member states. This Act underscores preventive regulation rather than the use of sanctions after the harm has been done. The EU AI Act officially came into effect on 1 August 2024, but its provisions will be implemented gradually—the Regulation (EU) of the European Parliament and the Council of 13 June 2024 has 144 pages.

After 44 pages of term definitions and an outline of its scope, the law repeats its aims, purpose, and definitions. Then, it describes prohibited AI practices, high-risk AI systems, and the actions intended to deal with the former and manage the latter. 'The Act assigns applications of AI to three risk categories. First, applications and systems that create an **unacceptable risk**, such as government-run social scoring of the type used in China, are banned. Second, **high-risk applications**, such as a CV-scanning tool that ranks job applicants, are subject to specific legal requirements. Lastly, applications not explicitly banned or listed as high-risk are largely left unregulated' (https://artificialintelligenceact.eu/).

I have excerpted the following paragraphs from an EU page (https://digital-strategy.ec.europa.eu/en/policies/regulatory-framework-ai) to provide a brief summary of this piece of legislation: 'The AI Act defines four levels of risk for AI systems:

Unacceptable Risk

All AI systems considered a clear threat to the safety, livelihoods and rights of people are banned. The **AI Act prohibits eight practices**, namely:

1. harmful AI-based manipulation and deception
2. harmful AI-based exploitation of vulnerabilities
3. social scoring
4. Individual criminal offence risk assessment or prediction
5. untargeted scraping of the internet or CCTV material to create or expand facial recognition databases
6. emotion recognition in workplaces and education institutions
7. biometric categorisation to deduce certain protected characteristics
8. real-time remote biometric identification for law enforcement purposes in publicly accessible spaces

High Risk

AI use cases that can pose serious risks to health, safety or fundamental rights are classified as high-risk. These **high-risk** use-cases include:

- AI safety components in critical infrastructures (e.g., transport), the failure of which could put the life and health of citizens at risk
- AI solutions used in education institutions, that may determine the access to education and course of someone's professional life (e.g., scoring of exams).

- AI-based safety components of products (e.g., AI application in robot-assisted surgery)
- AI tools for employment, management of workers and access to self-employment (e.g., CV-sorting software for recruitment)
- Certain AI use-cases utilized to give access to essential private and public services (e.g., credit scoring denying citizens opportunity to obtain a loan)
- AI systems used for remote biometric identification, emotion recognition and biometric categorisation (e.g., AI system to retroactively identify a shoplifter)
- AI use-cases in law enforcement that may interfere with people's fundamental rights (e.g., evaluation of the reliability of evidence)
- AI use-cases in migration, asylum and border control management (e.g., automated examination of visa applications)
- AI solutions used in the administration of justice and democratic processes (e.g., AI solutions to prepare court rulings)

High-risk AI systems are subject to **strict obligations** before they can be put on the market:

- adequate risk assessment and mitigation systems
- high-quality of the datasets feeding the system to minimize risks of discriminatory outcomes
- logging of activity to ensure traceability of results
- detailed documentation providing all information necessary on the system and its purpose for authorities to assess its compliance
- clear and adequate information to the deployer
- appropriate human oversight measures
- high level of robustness, cybersecurity and accuracy

Limited Risk

This refers to the risks associated with a need for transparency around the use of AI. The AI Act introduces specific disclosure obligations to ensure that humans are informed when necessary to preserve trust. For instance, when using AI systems such as chatbots, humans should be made aware that they are interacting with a machine so they can take an informed decision.

Moreover, providers of generative AI have to ensure that AI-generated content is identifiable. On top of that, certain AI-generated content should be clearly and visibly labelled, namely deep fakes and text published with the purpose to inform the public on matters of public interest.

Minimal or No Risk
The AI Act does not introduce rules for AI that is deemed minimal or no risk. The vast majority of AI systems currently used in the EU fall into this category. This includes applications such as AI-enabled video games or spam filters.

A Solution for the Trustworthy Use of Large AI Models
General-purpose AI (GPAI) models can perform a wide range of tasks and are becoming the basis for many AI systems in the EU. Some of these models could carry systemic risks if they are very capable or widely used. To ensure safe and trustworthy AI, the AI Act puts in place rules for providers of such models. This includes transparency and copyright-related rules. For models that may carry systemic risks, providers should assess and mitigate these risks.'

This law has sparked controversy and opposition, fuelling calls to make compliance easier and more feasible by avoiding unnecessary bureaucracy, clarifying legal uncertainties, and providing more funding for AI research. Indeed, it is always hard to regulate unfolding developments like AI. As Carlos Montemayor states, 'our current AI technology presents unique challenges, such as higher unpredictability and lack of explainable deliverables, but it still has the status of a complex machine that needs to be inspected and legally regulated' (2023, p. 204). Let us take a look at the codes of ethics before delving into the difficulties hindering AI accountability.

Codes of Ethics

Metzinger (EPRS, p. 28) argues that we need both a global AI Charter and a code of ethical conduct. 'Any AI Global Charter, or its European precursor, should always be complemented by a concrete Code of Ethical Conduct guiding researchers in their practical day-to-day work' (EPRS, p. 31), he notes.

The G7 has proposed 11 'guiding principles' for these codes (www.ey.com/en_gl/insights/ai/g7-ai-principles-and-code-of-conduct), which I am quoting below.

1. 'Take appropriate measures throughout the development of advanced AI systems, including prior to and throughout their deployment and placement on the market, to identify, evaluate and mitigate risks across the AI lifecycle.
2. Identify, report and mitigate patterns of misuse after deployment, including placement on the market.
3. Publicly report advanced AI systems' capabilities, limitations and domains of appropriate and inappropriate use in order to help ensure sufficient and ongoing transparency, and thereby contribute to increased accountability.
4. Work towards responsible information-sharing and reporting of incidents among organizations developing advanced AI systems, including industry, governments, civil society and academia.
5. Develop, implement, and disclose AI governance and risk management policies, grounded in a risk-based approach—including privacy policies and mitigation measures, in particular for organizations developing advanced AI systems.
6. Invest in and implement robust security controls, including physical security, cybersecurity, and insider-threat safeguards across the AI lifecycle.
7. Develop and deploy reliable content authentication and provenance mechanisms, where technically feasible, including watermarking or other techniques to enable users to identify AI-generated content.
8. Prioritize research to mitigate societal safety and security risks and prioritize investment in effective mitigation measures.
9. Prioritize the development of advanced AI systems to address the world's greatest challenges, such as the climate crisis, global health, and education.
10. Advance the development and adoption of international technical standards.
11. Implement appropriate data input measures and protections for personal data and intellectual property.'

Most AI companies have already adopted codes of ethics. For example, the French consulting company Capgemini's code of ethics states (https://www.capgemini.com/wp-content/uploads/2021/06/Capgemini_Code_of_Ethics_for_AI_2021_EN.pdf):

> At Capgemini, we believe that human ethical values should never be undermined by the uses made of AI by business. We want AI solutions to be human-centric, which we define as follows:
>
> 1. AI with carefully delimited impact—designed for human benefit, with a clearly defined purpose setting out what the solution will deliver, to whom.
> 2. Sustainable AI—developed mindful of each stakeholder, to benefit the environment and all present and future members of our ecosystem, human and non-human alike, and to address pressing challenges such as climate change, CO_2 reduction, health improvement, and sustainable food production.
> 3. Fair AI—produced by diverse teams using sound data for unbiased outcomes and the inclusion of all individuals and population groups.
> 4. Transparent and explainable AI—with outcomes that can be understood, traced, and audited, as appropriate.
> 5. Controllable AI with clear accountability—enabling humans to make more informed choices and keep the last say.
> 6. Robust and safe AI—including fallback plans where needed.
> 7. AI respectful of privacy and data protection—considering data privacy and security from the design phase, for data usage that is secure, and legally compliant with privacy regulations.

This quote comes from a brief summary of the company's code of ethics. While codes of ethics are usually very long and detailed, this excerpt shows the primary issues typically addressed by those codes. These are very useful instruments to try to instil a moral ethos in AI-related business practices.

Accountability

Novelli et al. (2024) have elaborated a comprehensive theory of accountability and have applied it to the AI system. They view accountability as cornerstone requirement for AI, and thus we need a clear definition of its nature, scope, and proceedings. The core of accountability lies in the obligation to inform and justify one's behaviour to an authority—it involves a relation based on 'answerability' (2024, p. 1872). This notion implies the recognition of an authority, an 'interrogation' (the agent is exposed to the scrutiny of an external forum), and a limitation of the agent's power. Novelli et al. list seven features of accountability: context (the field for an accountability relation), range (the about what), the accountable agent, the forum that examines the agent's actions, standard or principles (the rules used to judge the agent's actions), process, and implications. They also define the goals of accountability: compliance (alignment with ethical and legal standards), reporting relevant information, overseeing the agent's behaviour, and enforcement of applicable sanctions, prohibitions, etc.

The application of these concepts to AI raises some problems. AI system's opaque processes hamper the search for the underlying causes of its unpredictable, unintended, and potentially harmful outcomes. There are just too 'many hands' at work in AI processes and too 'many eyes' doing the oversight. Goal choices depend on whether the accountability pursued is proactive (ex-ante) or reactive (ex-post), and typically there is a combination of both. Proactive accountability focuses on generating the desired outcomes and avoiding the undesirable consequences by identifying them before they happen. Reactive accountability, in turn, deals with the rewards or sanctions once the outcomes unfold and become clear.

Concerning accountability in the 'computerized society,' Helen Nissenbaum (1996) highlighted the following issues:

- Again, the problem of 'many hands': While traditional ethics assigns responsibility to an individual when harm occurs, this is virtually impossible in the case of machine learning techniques, which are based on previous human-algorithm interactions, consistently updated by a number of engineers and incorporated to other algorithms that have their own creators and data sets. Given all these interactions, no single human being can be held solely accountable for effects of actions driven by algorithms. There is a shared responsibility that requires a sound and shared personal ethics.
- Bugs (virus): 'Software errors including modelling design and coding errors' (1996, p. 32) are viewed as inevitable in new developments, which means that programmers, engineers, and designers cannot be held accountable for them.
- 'It's the computer's fault': Using computers as scapegoats and blaming them for any harm done hinders accountability.
- Software ownership without liability: Software is usually provided 'as is,' with no warranty of any kind.

Cooper et al. (2022) revisit these four barriers to accountability in relation to algorithmic systems—i.e., artificial intelligence or machine learning. They conclude that these new systems heighten the hurdles for accountability and discuss ways to do away with these hindrances, such as establishing institutional frameworks to identify accountable parties.

In sum, accountability looks like a very difficult task when it comes to dealing with AI systems. Their opaqueness, autonomy, and complexity make it very hard to assign responsibilities to individual agents. However, resolute efforts must be geared towards determining the specific responsibilities of the multiple parties at play, on a case-by-case basis. A careful examination of AI processes by AI companies, developers, and users, followed by transparent reporting, seem to be necessary for appropriate accountability. We need an explainable AI system, and it is with this in mind that Ben Shneiderman presented a set of proposals to build and use a 'human-centred AI' (2022).

Conclusion

This chapter has surveyed Aristotle's ideas about the role of the State in securing people's well-being, with education and law promoting citizens' virtues and happiness, before exploring the regulations—declarations, laws, and codes of ethics—that have been devised to address AI's development and potential risks. These tools are useful to create an ethical atmosphere for AI systems, but some hurdles seem to curtail the existence of effective accountability in this field, given AI's autonomy, opacity, and the multiple actors involved in its development. The next chapter will look into the possibility of building 'moral machines' to reinforce AI's focus on human and societal good.

References

Aristotle. (1954). *Nicomachean Ethics* (Sir W. D. Ross, Trans. and Intro.). Oxford University Press.

Aristotle. (1958). *Politics* (E. Barker, Ed. and Trans.), Oxford University Press.

Aristotle. (1941). *The Basic Works of Aristotle* (R. McKeon, Ed. and with an Intro). Random House, New York (reprint of the translations prepared under the editorship of W. D. Ross, Oxford University Press).

Chesterman, S. (2021). *We, the Robots. Regulating Artificial Intelligence and the Limits of the Law*. Cambridge University Press.

Cooper, A. F., Moss, E., Laufer, B., & Nissenbaum, H. (2022). Accountability in an Algorithmic Society: Relationality, Responsibility, and Robustness in

Machine Learning. In FAccT '22: Proceedings of the 2022 ACM Conference on Fairness, Accountability, and Transparency, pp. 864–876.
Crespo, R. F. (2007). "Practical Comparability" and Ends in Economics. *Journal of Economic Methodology, 14*(3), 371–393.
Ekbia, H., Mattioli, M.,& Kouper, I., & Ghazi, A. (2015). Big Data, Bigger Dilemmas: A Critical Review. *Journal of the Association for Information, Science and Technology*. Retrieved October 23, 2024, from https://doi.org/10.1002/asi.23294
EPRS (European Parliamentary Research Service). (2018). Should We Fear Artificial Intelligence? Retrieved February 23, 2025, from www.europarl.europa.eu/RegData/etudes/IDAN/2018/614547/EPRS_IDA(2018)614547_EN.
Montemayor, C. (2023). *The Prospect of a Humanitarian Artificial Intelligence. Agency and Value Alignment*. Bloomsbury Academic.
Nida-Rümelin, J., & Staudacher, K. (2024). Philosophical Foundations of Digital Humanism. In H. Werthner, C. Ghezzi, J. Kramer, J. Nida-Rümelin, B. Nuseibeh, E. Prem, & A. Stanger (Eds.), *Introduction to Digital Humanism. A Textbook* (pp. 17–30). Springer.
Nida-Rümelin, J., & Weidenfeld, N. (2022). *Digital Humanism. For a Humane Transformation of Democracy, Economy and Culture in the Digital Age*. Springer.
Nida-Rümelin, J., & Winter, D. (2024). Humanism and Enlightenment. In H. Werthner, C. Ghezzi, J. Kramer, J. Nida-Rümelin, B. Nuseibeh, E. Prem, & A. Stanger (Eds.), *Introduction to Digital Humanism. A Textbook* (pp. 3–16). Springer.
Nissenbaum, H. (1996). Accountability in a Computerized Society. *Science and Engineering Ethics, 2*, 25–42.
Novelli, C., Taddeo, M., & Floridi, L. (2024). Accountability in Artificial Intelligence: What It Is and How It Works. *AI & Society, 39*, 1871–1882.
Nussbaum, M. C. (1987). Nature, Function, and Capability: Aristotle on Political Distribution, *WIDER Working Paper* 31, Helsinki.
Nussbaum, M. C. (1990). Aristotelian Social Democracy. In R. Bruce Douglass, G. M. Mara, & H. S. Richardson (Eds.), *Liberalism and the Good* (pp. 203–252). Routledge.
Shneiderman, B. (2022). *Human-Centered AI*. Oxford University Press.
Werthner, H. (2024). Digital Transformation, Digital Humanism: What Needs to Be Done. In H. Werthner, C. Ghezzi, J. Kramer, J. Nida-Rümelin, B. Nuseibeh, E. Prem, & A. Stanger (Eds.), *Introduction to Digital Humanism. A Textbook* (pp. 115–132). Springer.

Werthner, H., Ghezzi, C., Kramer, J., Nida-Rümelin, J., Nuseibeh, B., Prem, E., & Stanger, A. (Eds.). (2024). *Introduction to Digital Humanism. A Textbook*. Springer.

Werthner, H., Prem, E., Lee, E. A., & Ghezzi, C. (Eds.). (2022). *Perspectives on Digital Humanism*. Springer.

CHAPTER 8

'Moral Machines'?

Abstract This chapter tries to establish whether it is possible—and preferable—to create 'moral machines.' First, it describes the advantages and drawbacks typically associated with 'moral machines,' and, then, it introduces Aristotle's notion of and conditions for morality. Also, a summary of the ontological analysis of AI machines carried out in Chap. 3 is cited, as it seems fitting to determine whether they can indeed be moral. The conclusion is that the ontological condition of artifacts prevents machines from having the internal capabilities required to act with an end in sight, and, therefore, they cannot act morally. They lack a conscience, free will, and sentience, and, thus, it is impossible for moral machines to exist. However, as AI agents' actions actually have moral consequences, we must provide the means to keep them from doing harm. Alas, given the nature of ethics, the last word must be human. AI alone should not make final decisions, because its very nature prevents it from factoring in all the circumstances that moral appraisals take into account.

Keywords Moral Machines • The Ontology of Artifacts • Aristotle on Morality • Freedom • AI's Ethical Consequences • Harm Prevention

The previous chapter introduced declarations, regulations, codes of ethics, and accountability devices aimed at orienting and assuring a development

R. F. Crespo, *A Classical Philosophical Approach to Artificial Intelligence*, https://doi.org/10.1007/978-3-032-18462-7_8

of AI that promotes the human and social good. Their efficacy, however, remains questionable. This chapter will analyse the possibility of creating 'moral machines' as a means to reinforce the effectiveness of those mechanisms.

The term 'moral machine'—or, more precisely, the phrase 'machine ethics'—was coined by Mitchell Waldrop in his 1987 article 'A Question of Responsibility' (1987). This piece starts by quoting the three ethical principles famously posited by Isaac Asimov (1942), also known as his 'Three Laws of Robotics':

1. A robot may not injure a human being or, through inaction, allow a human being to come to harm.
2. A robot must obey orders given it by human beings except where such orders would conflict with the First Law.
3. A robot must protect its own existence as long as such protection does not conflict with the First or Second Law.

Then, after describing how AI works and the challenges it poses, Waldrop concludes (1987, p. 32):

> […] intelligent machines will embody values, assumptions, and purposes, whether their programmers consciously intend them to or not. Thus, as computers and robots become more and more intelligent, it becomes imperative that we think carefully and explicitly about what those built-in values are. Perhaps what we need is, in fact, a theory and practice of machine ethics, in the spirit of Asimov's three laws of robotics.

Nonetheless, this topic did not gather momentum until 2004, when a paper written by Michael Anderson, Susan Leigh Anderson, and Chris Armen was presented at the 'AAAI-04 Workshop on Agent Organizations: Theory and Practice.' Since that massive boom, the number of books and articles published on this matter has grown immensely. A seminal paper on it was penned by James H. Moor (2006, 2009), who listed several moral machine agents, including ethical impact, implicit, explicit, and full ethical agents. As he explained (2009),

> **Ethical impact agents** are agents whose actions have ethical consequences, whether intended or not.

> **Implicit ethical agents** are agents that have ethical considerations built into (i.e., implicit in) their design. Typically, these are safety or security considerations.
>
> **Explicit ethical agents** are agents that can identify and process ethical information about a variety of situations and make sensitive determinations about what should be done. When ethical principles are in conflict, these robots can work out reasonable resolutions. Explicit ethical agents are the kind of agents that can be thought of as acting *from* ethics, not merely *according to* ethics.
>
> ***Full* ethical agents** have those central metaphysical features that we usually attribute to ethical agents like *us*—features such as consciousness, intentionality, and free will. Ordinary adult humans are the prime example of full ethical agents.

Current discussions focus on whether it is convenient or not to create explicit ethical agents, because the existence of full ethical agents implies that 'Artificial General Intelligence' (AGI) is possible, which seems doubtful today.

There are arguments for and against explicit moral machines. Concerning the pros, Moor (2006, p. 20) affirms that 'Confronted with a complex problem requiring fast decisions, computers might be more competent than humans.' He adds:

> These decisions could be ethical in that they would determine who would live and who would die. Some might say that only humans should make such decisions, but if (and of course this is a big assumption) computer decision making could routinely save more lives in such situations than human decision making, we might have a good ethical basis for letting computers make the decisions.

The implicit ethical theory in this argument is utilitarianism. Moor provides a three-prong rationale in support of explicit ethical agents (2006, p. 21):

> I can offer at least three reasons why it's important to work on machine ethics in the sense of developing explicit ethical agents:
>
> - Ethics is important. We want machines to treat us well.

- Because machines are becoming more sophisticated and make our lives more enjoyable, future machines will likely have increased control and autonomy to do this. More powerful machines need more powerful machine ethics.

- Programming or teaching a machine to act ethically will help us better understand ethics.

These are the pros upheld by people in favour of moral explicit machines. The key argument is that AI agents' actions actually have ethical consequences and thus we should imbue them with ethical criteria. Also, moral machine advocates think that these machines could behave even more ethically than humans because they are impartial—not impaired by potential human biases (Anderson & Anderson, 2007, 2010, p. 77).[1]

Exploring the source of human beings' morality and its differences with machines may clarify these matters. In a nutshell, can a machine as AI be moral? I will first introduce Aristotle's thinking about morality, before analysing whether AI can fit in with the nature and meet the conditions of morality.

Aristotle on Morality

Aristotle separated two parts of the rational [*lógon*] part of the soul: the scientific and the calculative aspects. The former deals with things that are invariable, while the latter (called calculative [*logistikon*]) handles things that can be variable (*NE* VI, 1, 1139a1–17). In turn, the proper object of the former is truth, and that of the latter is truth as pertaining to a rightful desire, which is the good [*tò eû*] (*Nicomachean Ethics* VI, 2, 1139a29–30).

In *De Anima* III, 9, 432b5, Aristotle stated that '*én te tô logistikô gar he boúlesis gínetai*'—that is, 'there is will in the calculative part.' The mind never moves without the appetite [*oréxis*] (and will [*boúlesis*] is a form of appetite) (*De Anima* III, 10, 433a23–24). He continued, 'Mind [*noûs*] is always right; but appetite and imagination may be right or wrong. Thus the object of appetite always produces movement, but this may be either the real or the apparent good' (*De Anima* III, 10, 433a27–30). This is the

[1] Aimee van Wynsbergue and Scott Robbins (2019) criticise the arguments in favour of moral machines—inevitability, prevention of harm, complexity of AI agents, providing public trust, preventing immoral use, and better performance than humans.

condition of practical good [*pràkton agathón*], that it can be otherwise (III, 10, 433a30–31). In other words, the will can choose to follow an apparent good, not the real good that the right practical reason presents to it. This ability of the will is what we call inner freedom—the choice to pursue the good or an apparent good—and it is a necessary condition for morality. Only free beings can behave morally well or not.[2] Another condition is that we have a pre-established natural end to pursue (and decide to pursue it or not), and this end depends on human beings' nature. The content of this end has been proposed by Aristotle in *Nicomachean Ethics*, his book on the virtues necessary to have a good life as defined by practical reason.

The human end and people's character do not determine their actions univocally. For example, in *Nicomachean Ethics,* Aristotle wrote that passion 'does not master to the state of making [a person] ready to believe that he ought to pursue such pleasures without reserve' (VII, 8, 1151a20–23; see also VII, 7, 1150b30—'any incontinent man is likely to repent'—and 1151a 20–23). A voluntary act, he stated, 'would seem to be that of which the moving principle is in the agent himself, he being aware of the particular circumstances of the action' (III, 1, 1111a20–21). Another passage, in III, 5, 1113b7–8, underscores this notion: 'For where it is in our power to act it is also in our power not to act, and *vice versa.*'[3]

Another condition for freedom is indeterminism, and Aristotle regarded necessity not as absolute but as hypothetical. Events are generated by a conditional convergence of causes that do not always occur simultaneously. He noted that 'some cases, moreover, we find that, at least, for the most part, and commonly, tend in a certain direction, and yet they may issue at times in the other or rarer direction' (*On Interpretation* IX, 19a 20–3), and 'those things that are not uninterruptedly actual exhibit a potentiality, that is, a may be or a may not be. If such things may be or may

[2] For Aristotle, causes are internal or essential powers or capacities triggered (or not) by an external 'efficient cause' moved by the force or attraction exerted on it by a 'final cause.' Instead, the final cause does not exist because it is desired or intended by the efficient cause. Aristotle viewed desire as necessary to trigger an action, but desires can differ from goals defined by human nature and practical reason—reasons might differ from desires.

[3] This edition by Ross continues with a rather poorly translated passage—'*kaì en oîs tò mé, kaì tò naí*'—that others have translated as 'and when saying no does so, saying yes does too.' The meaning and consequences of freedom for Aristotle have been discussed somewhat extensively. See, for example, Sarah Broadie (1991, Chapter 3) and Suzanne Bobzien (2014) for different views on this.

not be, events may take place or not' (*On Interpretation* IX, 19a 10–3). Aristotle was neither a determinist nor—consequently—a fatalist.[4] The reality of contingency expressed in these passages allows for free actions. In the discussion about the compatibility or lack thereof between Aristotle's voluntary actions and determinism, I fall on the incompatibilist side, because the whole spirit of *Nicomachean Ethics* supports the notion of acting against possible deterministic conditions, and Aristotle's defence of indeterminism is clear in the previously cited passages (see Broadie, 1991, Chapter 3).

Yet, contingency is not sufficient: we need a voluntary intention that Aristotle also noted. As Elizabeth Anscombe ([1971] 1981, p. 146) asserts, 'The truth of physical indeterminism is thus indispensable if we are to make anything of the claim to freedom. But certainly it is insufficient. The physically undetermined is not thereby "free." For freedom at least involves the power of acting according to an idea, and no such thing is ascribed to whatever is the subject (what would be the relevant subject?) of unpredetermination in indeterministic physics.' This power is ascribed to the human being.

Aristotle also elaborated on the voluntary act (*boulesis*) in the *Eudemian Ethics* II, 8: 'what is in conformity to one's will is not involuntary; rather, everything that is willed is also voluntary' (II, 8, 1224a1–3). And back in the *Nicomachean Ethics,* he sentenced that 'we are masters of our actions from the beginning to the end' (III, 6, 1114b30–1115a1). Thus, Aristotle clearly believed that we human beings generate our voluntary actions and can do the opposite of what we actually do. We are influenced by our character, but our character is also shaped by voluntary actions, and we can even act against it.[5]

In sum, the conditions for morality according to Aristotle's works are physical indeterminism, the ability to act voluntarily and freely, and the existence of an end, i.e., the good life (*eû zên*). The 'construction' of morality hinges on our voluntary freedom to develop a virtuous character. To explore whether these conditions for morality are present in AI, I briefly turn to the ontology of AI expounded in Chap. 4.

[4] In *Physics* (II, 9) he also stated that necessity is hypothetical in the material realm. On this, see my book Crespo 2013, pp. 23–24; Auguste Mansion's classic on *Physics,* 1913, pp. 171–178, and Richard Sorabji (1980, Chapter 9).

[5] For a defence of human freedom in the context of AI, see Julian Nida-Rümelin and Nathalie Weidenfeld (2022, Chapters 4 and 5).

Can AI Be Moral?

Chapter 4 on ontology focuses on the nature of AI, an abstract artifact. To start that analysis, I laid forth the three traits that characterize artefacts: having been produced intentionally, involving a modification of materials, and featuring a purpose. It might come in handy to revisit David Wiggins' quotation about artefacts, setting them apart from natural substances:

> Artefacts are things to be encountered in nature, and subject to its laws. But they need not be collected and classified together as this or that artefact in virtue of internal resemblances or scientific resemblances. Artefacts are collected up not by reference to a theoretically hypothesized common constitution but under functional descriptions that are precisely indifferent to specific constitution and particular mode of interaction with the environment. (Wiggins, 2001, p. 87)

In other words, while natural substances are classified by their nature, artefacts are classified by their function. Another difference is that natural substances, Wiggins states, have natural propensities that constitute their foundation (2001, p. 88). He quotes Aristotle on this point (2001, p. 81):

> Of things that exist, some exist by nature, some from other causes. 'By nature' the animals and their parts exist, and the plants and the simple bodies (earth, fire, air, water)—for we say that these and the like exist 'by nature.' All the things mentioned present a feature in which they differ from things which are not constituted by nature. Each of them has within itself a principle of motion and of stationariness (in respect of place, or of growth and decrease, or by way of alteration). On the other hand, a bed and a coat and anything else of that sort, qua receiving these designations i.e., in so far as they are products of art–have no innate impulse to change. But in so far as they happen to be composed of stone or of earth or of a mixture of the two, they do have such an impulse, and just to that extent which seems to indicate that nature is a source or cause of being moved and of being at rest in that to which it belongs primarily, in virtue of itself and not in virtue of a concomitant attribute. I say 'not in virtue of a concomitant attribute,' because (for instance) a man who is a doctor might cure himself. Nevertheless it is not in so far as he is a patient that he possesses the art of medicine: it merely has happened that the same man is doctor and patient—and that is why these attributes are not always found together. So it is with all other artificial products. None of them has in itself the source of its own production. (*Physics* II 1 192b 8–28)

The internal adequate disposition of the natural parts of artefacts enables them to operate. While Aristotle believed that natural substances have an internal principle of action or power, artefacts receive this principle of motion from their producers thanks to the appropriate combination of natural components. Jacobs explains:

> [...] the craftsman cannot create a real substance with a real essence; rather, he uses already existing substances as instruments to an end, and so the artefact does not have real existential unity. It has an intentional unity arising from that conceived form accidentally uniting those substances; it is this intentionality that is the extrinsic source of unity for the artefact and so acts as the essence of the artefact. (2019, p. 153)

Given their free will, human beings can act (or not) according to their natural end, and the realm of morality stems from this very fact. Instead, as artefacts have neither an internal principle of motion tending to an end nor free will, their end is imprinted by human beings, who are responsible for their morality. That is, because of AI's nature, it is ontologically impossible for it to be moral. The autonomous ontogenetic nature of AI described in Chap. 4 should not be construed as free will—its processes are merely automatic. Referring to the random element of digital computers, Alan Turing comments, 'Sometimes such a machine is described as having free will (though I would not use this phrase myself)' (1950, p. 438). In 'The Philosophy of AI and the AI of Philosophy,' John McCarthy, 'the father of AI,' takes a compatibilist stance between free will and determinism. Only thus, he affirms, we can speak of 'a kind of *free will* in a logical robot' (2006, p. 14, italics in the original). However, we have adopted an incompatibilist position.

The AI agent is not an agent in the sense described in the previous section, i.e., a human moral agent, but an IA that can perform complex tasks autonomously is a 'non-human agent.' This AI's autonomy is essentially different from human freedom, and using the term 'agent' to refer to AI is equivocal because human agency implies freedom and intentionality, which AI lacks. As a result, AI is also essentially different and cannot be moral.[6]

[6] Fritz et al. (2020) argue at length against the notion of agency and of moral agency applied to AI.

There are other arguments that debunk the notion of explicit ethical or moral agents. First, ethics is not a matter of statistics. A machine learning algorithm can extract the most probable ethical behaviour for a specific scenario from a number of ethical situations that it analyses. Yet, ethics is not determined by probability, and morality is not codifiable. Génova et al. explain, 'Ethics cannot be based on a majority opinion about right and wrong or on a rigid code of conduct. [...] The attempt to formalize ethics in a set of rules misses the point that a person is not only an instance of a case, but a unique and unrepeatable being' (2022, p. 71). Ethics remains a practical wisdom task that requires prudence, as well as some basic principles that call for a specific interpretation.

Second, machines lack emotions, and thus they cannot assess the feelings of people affected by specific ethical situations. Carissa Véliz (2021) states that algorithms lack *sentience*, a capability for a subjective experience of pleasure and pain that proves necessary for moral understanding to be morally responsible. For other authors like Kenneth E. Himma, morality presupposes having a conscience. He argues that 'only a conscious being can be an agent' (2009, p. 28). Two capabilities are necessary and jointly sufficient for moral agency: the ability to freely choose how to act and the capacity to know the difference between right and wrong—moral reasoning. Anuj Puri (2020) also points out that free will, consciousness, and moral intentionality are necessary for a moral agent, and AI lacks these traits.

Third, it proves challenging to determine the ethical theory and values needed to program AI agents. It seems that a combination of deontology and virtue ethics would be the best option, but the question remains as to how to program this. Additionally, as Iason Gabriel (2020, p. 412) asks,

> Behind each vision for ethically aligned AI sits a deeper question. How are we to decide which principles or objectives to encode in AI—and who has the right to make these decisions—given that we live in a pluralistic world that is full of competing conceptions of value? Is there a way to think about AI value alignment that avoids a situation in which some people simply impose their views on others?

What would a fair process to determine the values that should be adopted look like or entail? This remains an open question, as its answer hinges on AI's ability to behave morally. Based on its nature, is it even possible for AI to behave morally? I have argued that it is not.

Aimee van Wynsbergue and Scott Robbins (2019, pp. 724–725) make a distinction between 'being in a morally charged situation' and 'being delegated a moral role,' arguing that entrusting a moral capacity to AI is not necessary and can even prove harmful. These authors posit that 'while autonomous robots and AI can and should be used in morally salient contexts this need not require that the robot be endowed with ethical reasoning capabilities' (2019, p. 731). For them, the key concern about AI zeroes in on ensuring safety, but ethics cannot be reduced to that. As they see safety—not moral agency—at play here, they remark, 'One should not refer to moral machines, artificial moral agents, or ethical agents if the goal is really to create safe, reliable machines' (2019, p. 732). Amanda Sharkley (2020, pp. 290ff.) shares this perspective, stating that we should try to avoid putting robots in situations that involve ethical decisions and social context assessments (2020, p. 293).

Nonetheless, artefacts may ultimately have moral consequences or not, but this morality belongs to the human beings that build or use them. In sum, artificial entities are not moral, but their actions can have moral outcomes. However, the fact that artefacts can have moral outcomes does not require them to have been built as explicit or full ethical machines. The responsibility for their consequences lies with their builders and users, and it encompasses AI's autonomous developments, which were not purposefully designed by AI's creators. Almost a century ago, Russian philosopher Nicholas Berdiaeff asserted:

> It is not machinery, which is merely man's creation and consequently irresponsible, that is to be blamed, and it is unworthy to transfer responsibility from man to machine. Man alone is to blame for the awful power that threatens him. (1933, p. 50, 1972, p. 212)

That said, as AI agents actually do have moral consequences, measures must be taken to avoid possible harmful effects, for example, in the cases of self-driving vehicles, autonomous weapons, health choices, care of old people, potentially biased labour, or financial decisions, etc.

In addition, Thilo Hagendorff and David Danks (2023) note that some 'large language models' have been developed as 'morally informed AI systems.' They can have two sources: a bottom-up approach (picking up moral situations) and a bottom-down outlook (prescriptive rules derived from ethical theories). A possibly biased advice from the first source would be corrected by the bottom-down rules. However, Hagendorff & Danks,

2023, pp. 559–560) stress the difficulties involved in these developments and argue that, if rolled out, 'these systems should never be the sole arbiters of real-world situations' (Hagendorff & Danks, 2023, p. 559). These actions would amount to a useful first step, but they should not suffice: the final word must be human. Prescriptive bottom-down rules pose interpretation problems concerning the ethical character of specific behaviours in specific contexts or cultures.

Thus, people must pay attention to the evolution and consequences of AI and exercise reasonability: the time has come to go back to classical practical reasoning. Génova et al. underline that 'we need to recover value rationality beyond numbers, beyond logical and instrumental rationality' (2022, p. 80). In a later article (2023, p. 10), they suggest adopting virtue ethics, which focuses on being good people. They remark:

> Ethics is not an imitation game (Génova et al., 2022). Ethics is not about following a code of conduct like Asimov's Three Laws. But neither is ethics about imitating the behaviour of others. Learning by imitation is something very human, but if there is any ethical difference between doing A and doing B, that difference is not in the number of people who behave one way or another, but in the reasonableness and good will of their behaviour.

Clearly, this highlights the importance of human beings' presence and intervention. Julian Nida-Rümelin and Nathalie Weidenfeld stress the relevance of the human deliberation ability, which cannot be replaced by an optimization criterion: 'moral deliberation can only be done by human beings' (2022, p. 43).

Conclusion

The ontological condition of artefacts prevents AI agents from having the internal drive towards an end, and, therefore, they cannot act morally. Artefacts do not have a conscience, free will, or sentience, and, thus, it is impossible for moral machines to exist. Still, as AI agents actually have moral consequences, we must provide the means to prevent them from doing harm, especially in AI developments that could affect human safety, health, and dignity. This does not mean that they are moral machines, but it does underscore the need for safety measures, which will prove easier to incorporate into machines than to roll out in specific domains (see Susan Leigh Anderson, 2011, p. 25).

Recent AI developments that mirror human behaviour can be in fact ethically informed systems. As human behaviour is often unethical, imbuing AI with bottom-down basic ethical principles—like the Jewish commandments—might be useful. Human behaviour may be ethically wrong or biased, and ethical principles may include conflicts of principles. Indeed, ethical tenets can provide guidelines, but only as implicit ethical agents, according to Moor's classification, designed to avoid negative impacts for safety or security reasons (see Wilfield et al., 2019, p. 512). Also, if indeed ethical 'algorithms are imposed on the machine, the machine is in itself not an artificial moral agent, regardless of how well it performs' (Patrick Chisan Hew, 2014, p. 204). In this case, all we have is human ethics operating through a tool. Even in their defence of moral machines, Colin Allen, Wendell Wallach, and Iva Smit admit that they are dealing with 'artificial morality,' not genuine ethical agents that are free to act as they see fit (2011, p. 53).

As readers have probably realized by now, the topic is not easy. The boundaries between avoiding harmful effects and introducing ethical inputs are blurry, but it is clear that, given the nature of ethics, the final word must be human. As already mentioned in Chap. 2, Shannon Vallor's (2016) proposal for 'technomoral' virtues may help human beings to make these final choices.[7]

References

Allen, C., Wallach, W., & Smit, I. (2011). Why Machine Ethics? In M. Anderson & S. L. Anderson (Eds.), *Machine Ethics* (pp. 51–61). Cambridge University Press.

Anderson, M., Anderson, S. L., & Armen, C. (2004). Towards Machine Ethics, AAAI-04 Workshop on Agent Organizations: Theory and Practice.

Anderson, M., & Anderson, S. L. (2007). Machine Ethics: Creating an Ethical Intelligent Agent. *AI Magazine, 28*(4), 15–26.

Anderson, M., & Anderson, S. L. (2010, October). Robot be Good. *Scientific American, 303*, 72–77.

Anderson, S. L. (2011). Machine Metaethics. In M. Anderson & S. L. Anderson (Eds.), *Machine Ethics* (pp. 21–27). Cambridge University Press.

Anscombe, G. E. M. ([1971] 1981). Causality and Determination. In *The Collected Philosophical Papers of G.E.M. Anscombe*, Volume Two, *Metaphysics and the Philosophy of Mind*. Basil Blackwell.

[7] See also Thilo Hagendorff (2022).

Aristotle. *Physics.* Translation by R. P. Hardie and R. K. Gaye, Retrieved August 24, 2024, from https://classics.mit.edu/Aristotle/physics.html.

Aristotle. (1925). *Nicomachean Ethics* (W. D. Ross, Trans.). Oxford University Press.

Aristotle. (1984). *The Complete Works of Aristotle* (J. Barnes, Ed.). Princeton University Press.

Aristotle. (2011). *The Eudemian Ethics* (A. Kenny, Trans.). Oxford University Press.

Asimov, I. (1942). Rundaround. *Astounding Science Fiction, 29*(1), 94–103.

Berdiaeff, N. (1972). Man and Machine. In C. Mitcham and R. Mackey (eds.), *Philosophy and Technology. Readings in the Philosophical Problems of Technology* (pp. 203–213). The Free Press. French edition, *L'Homme et la Machine* (1933). Paris: Éditions "Je Sers".

Bobzien, S. (2014). Aristotle's *Nicomachean Ethics* 1113b7–8 and free choice. In R. S. Pierre Destrée & M. Zingano (Eds.), *What is Up to Us? Studies on Causality and Responsibility in Ancient Philosophy.* Academia Verlag.

Broadie, S. (1991). *Ethics with Aristotle.* Oxford University Press.

Crespo, R. F. (2013). *Theoretical and ¨Practical Reason in Economics.* Springer Verlag.

Fritz, A., Brandt, W., Gimpel, H., & Bayer, S. (2020). Moral Agency Without Responsibility? Analysis of Three Ethical Models of Human-computer Interaction in Times of Artificial Intelligence (AI). *De Ethica, 6*(1), 3–22.

Gabriel, I. (2020). Artificial Intelligence, Values, and Alignment. *Minds and Machines, 30*, 411–437.

Génova, G., Pelayo, V. M., & Martín, M. R. G. (2022, March). A Lesson from AI: Ethics Is Not an Imitation Game. *IEEE Technology and Society Magazine, 41*, 75–81.

Génova, G., Pelayo, V. M., & Martín, M. R. G. (2023). Machine Ethics: Do Androids Dream of Being Good People? *Science and Engineering Ethics, 29*, 10.

Hagendorff, T. (2022). A Virtue-Base Framework to Support Putting AI Ethics into Practice. *Philosophy and Technology, 35*, 55.

Hagendorff, T., & Danks, D. (2023). Ethical and Methodological Challenges in Building Morally Informed AI Systems. *AI and Ethics, 3*, 553–566.

Hew, P. C. (2014). Artificial Moral Agents are Infeasible with Foreseeable Technologies. *Ethics and Information Technology, 16*(3), 197–206.

Himma, K. E. (2009). Artificial Agency, Consciousness, and the Criteria for Moral Agency: What Properties Must an Artificial Agent Have to be Moral Agent? *Ethics and Information Technology, 11*, 19–29.

Jacobs, J. M. (2019). The Ontological Status of Artifacts. In C. Cullen & F. Harkins (Eds.), *The discovery of being & Thomas Aquinas: Philosophical and Theological Perspectives* (pp. 145–160). The Catholic University of America Press.

Mansion, A. (1913). *Introduction à la Physique Aristotélicienne.* Félix Alcan.

McCarthy, J. (2006). The Philosophy of AI and the AI of Philosophy. Retrieved August 2, 2024, from http://jmc.stanford.edu/articles/aiphil2.html.

Moor, J. H. (2006). The Nature, Importance, and Difficulty of Machine Ethics. *Intelligent Systems, IEEE, 21* (4), 18–21, reprinted in Michael Anderson and Susan Leigh Anderson (eds., 2011). *Machine Ethics* (pp. 13–20). Cambridge: Cambridge University Press.

Moor, J. H. (2009). Four Kinds of Ethical Robots. *Philosophy Now*, 72, Retrieved June 24, 2025, from https://philosophynow.org/issues/72/FourKindsofRobots.

Nida-Rümelin, J., & Weidenfeld, N. (2022). *Digital Humanism. For a Humane Transformation of Democracy, Economy and Culture in the Digital Age*. Springer.

Puri, A. (2020). Moral Imitation: Can an Algorithm Really be Ethical? *Rutgers Law Records, 48*(1), 47–57.

Sharkley, A. (2020). Can we Program or Train Robots to Be Good? *Ethics and Information Technology, 22*, 283–295. https://doi.org/10.1007/s10676-017-9425-5

Sorabji, R. (1980). *Necessity, Cause and Blame. Perspectives on Aristotle's Theory*. Duckworth.

Turing, A. M. (1950). Computing Machinery and Intelligence. *Mind, New Series, 59*(236), 433–460.

Vallor, S. (2016). *Technology and the Virtues. A Philosophical Guide to a Future Worth Wanting*. Oxford University Press.

van Wynsbergue, A., & Robbins, S. (2019). Critiquing the Reasons for Making Artificial Moral Agents. *Science Engineering Ethics, 25*, 719–735.

Véliz, C. (2021). Moral Zombies: Why Algorithms Are Not Moral Agents. *AI & Society, 36*, 487–497.

Waldrop, M. (1987). A Question of Responsibility. *AI Magazine, 8*(1), 28. https://doi.org/10.1609/aimag.v8i1.572

Wiggins, D. (2001). *Sameness and Substance Renewed*. Cambridge University Press.

Wilfield, A., Machael, K., Pitt, J., & Evers, V. (2019). Machine Ethics: The Design and Governance of Ethical AI and Autonomous Systems. *Proceedings of the IEEE, 107*(3), 509–517.

CHAPTER 9

Strengthening the Remedies

Abstract Chapter 7 described the declarations, regulations, codes of ethics, and accountability mechanisms crafted for AI systems over recent years. Those tools have contributed to instil an ethical ethos into the AI field, but we also need specific laws and policies to guide AI developments. However, multiple and powerful political and economic interests conspire against an effective regulation for AI, as companies and—albeit somewhat obscurely—countries race frantically to dominate the AI system. We thus need to strengthen the remedies already described. Drawing from Norbert Wiener, Louise Amoore, Shannon Vallor, and Thilo Hagendorff, this chapter argues that we need a radical change in our personal behaviour and the actions of political and economic institutions in order to make the AI system work in collaboration with human beings and societies. This shift primarily requires enforcing the principles proclaimed by recent AI declarations, legislations, and codes of ethics, enhancing them with a real ethical conduct driven by the cultivation of specific ethical virtues.

Keywords Wiener on technology • Amoore on 'Cloud Ethics' • Vallor on technomoral virtues

In Chap. 7, I described the declarations, regulations, codes of ethics, and accountability mechanisms crafted for AI systems over recent years. I truly

R. F. Crespo, *A Classical Philosophical Approach to Artificial Intelligence*, https://doi.org/10.1007/978-3-032-18462-7_9

believe that those declarations have contributed to instil an ethical ethos in the AI field. However, I also think that we need specific laws and regulations to guide AI developments. Thilo Hagendorff (2020) has analysed the impact of 22 ethical guidelines published by companies and institutions on human decision-making in the field of AI and machine learning, and he has come to the conclusion that they have no effect. AI's ethics, he states, lacks the means to reinforce its normative claims. Hagendorff quotes Ulrich Beck, stating that ethics nowadays 'plays the role of a bicycle brake on an intercontinental airplane' (1988, p. 194).

In the same chapter, I also described the EU's AI Act, copying its list of AI risks and the measures associated with every one of them. This law has marked a clearly relevant step, because it is the type of legislation that should be adopted by other countries. Given the AI system's nature, its governing rules cannot be adopted by isolated countries if they are to prove effective. AI's characteristics, speed, autonomy, and opacity call for the involvement of States, as Simon Chesterman (2021, p. xv) remarks. However, he also notes that 'those states at the forefront of AI development—China and the United States—are, for different reasons, among those wariest of international law and institutions constraining their economic development and political independence. The result is that the states with the greatest leverage to establish global norms on AI presently have the least interest in doing so' (2021, pp. xv–xvi).

Multiple and powerful political and economic interests conspire against an effective regulation for AI. We are witnessing a frantic race among companies and—albeit somewhat hidden—among countries as well to dominate the AI system (see Hagendorff, 2020, p. 107). We thus need to strengthen the remedies already described, once again relying on Shannon Vallor's metaphor of the AI system as a mirror of ourselves to understand that AI will do that what we do. Hence, 'AI won't enable a sustainable future without reformed political institutions and economic incentives' (Vallor, 2024, p. 9).

Drawing from Norbert Wiener, who pioneered cybernetics (though different, an antecedent of AI), Louise Amoore (2020), and Vallor (2016, 2024), this chapter will argue that the remedies outlined earlier are not enough: all stakeholders involved with algorithms, AI, and machine learning processes—including States, companies, designers, developers, and users—must share a profound ethical inclination. As stated in the introduction, the effectiveness of the proposed measures has been called into question.

Terrell Bynum (2004) refers to Wiener's suggestions to cope with the challenges posed by new technologies:

> Wiener discusses what would count as a good human life. To have a good life, human beings must live in a society where 'the great human values which man possesses' (1954, p. 52) are nurtured; and this can only be achieved, he said, in a society that upholds the 'great principles of justice' (1954,, p. 106). In Chapter VI of *The Human Use of Human Beings* he stated those principles, although he did not give them names.[1] For the sake of clarity and ease of remembering them, let us attach names to Wiener's own definitions:
> The Principle of Freedom: Justice requires 'the liberty of each human being to develop in his freedom the full measure of the human possibilities embodied in him.' (1954,, p. 105)
> The Principle of Equality: Justice requires 'the equality by which what is just for A and B remains just when the positions of A and B are interchanged.' (1954,, p. 106)
> The Principle of Benevolence: Justice requires 'a good will between man and man that knows no limits short of those of humanity itself.' (1954,, p. 106)
> Wiener considered humans to be fundamentally social beings who can reach their full potential only by active participation in a community of similar beings. For a good human life, therefore, society is indispensable. But it is possible for a society to be oppressive and despotic in ways that limit or even stifle individual freedom; so Wiener added a fourth principle of justice, which we can appropriately call 'The Principle of Minimum Infringement of Freedom': (Wiener himself did not give it a name.)
> The Principle of Minimum Infringement of Freedom: 'What compulsion the very existence of the community and the state may demand must be exercised in such a way as to produce no unnecessary infringement of freedom.' (1954,, p. 106)

Bynum also affirms referring to Wiener's thought (2004, p. 68):

> all those in society who must cope ethically with the introduction of ICT [information and communication technology]—whether they are public policy makers, ICT professionals, business people, workers, teachers, parents, or others —can and should engage in Information Ethics by helping to integrate ICT into society in ways that are socially and ethically good.

[1] Wiener specifically states that the best words to express the requirements for the existence of justice are those of the French Revolution: 'Liberté, Egalité, Fraternité' (1954, p. 105).

> Information Ethics, understood in this very broad way, is too vast and too important to be left only to academics or to ICT professionals. This was clear to Wiener, who especially challenged government officials, business leaders, and public policy makers to wake up and begin to address the 'good and evil' implications of the coming information society.

In a later work, Wiener himself wrote about a sin 'which consists of using the magic of modern automatization to further personal profit' (1966, p. 52), a sin committed by 'gadget worshipers,' with their 'slogans of free enterprise and the profit-motive economy' (1966, p. 53).[2]

Wiener's ethical ideas are often included in codes of ethics. Nonetheless, as Madary and Metzinger (2016, p. 12) write in their ethical code of conduct, which makes recommendations for good scientific practice in virtual reality, 'Scientists must understand that following a code of ethics is not the same as being ethical. A domain-specific ethics code, however consistent, developed and fine-grained future versions of it may be, can never function as a substitute for ethical reasoning itself' (2016, p. 31).

Codes of ethics do not suffice if people are not ethical. Amoore, quoting Michel Foucault and Gilles Deleuze, separates codes of ethics from ethical practice (2020, p. 7; see also endnote 16, page 174), noting (2020, p. 165):

> In this book I differentiate the search for an encoded ethical framework for algorithms from an ethicopolitics that is already present within the algorithm's arrangements. In so doing, I draw a distinction between a moral code of prohibition and permission (what algorithms may or may not do) and an ethics of the orientation one has to oneself and to others (how algorithms are generating relations of self to self, and of self to others) [...]. Our ethicopolitical relations do not follow strictly from some already formulated moral code, for we never have the security of these grounds from which to act.

Referring to Isabelle Stengers, Amoore clarifies that the speculative ambition is ethical in the sense of ethos or habit (2020, p. 165), and quoting Simon Critchley, she highlights the relational character of ethics (2020,

[2] Martin Heidegger also takes a deeply critical view of the prevalence of cybernetics. He wrote, 'the world as represented in cybernetic terms abolishes the difference between automatic machines and living beings. The cybernetic blueprint of the world [...] makes possible a complete homogeneous—and in this sense universal—calculability, that is, the absolute controllability of both the animate and the inanimate world' (2013, p. 123). He would have referred to AI in these terms if he had known it.

p. 171). Amoore takes for granted the biased characteristic of algorithms and views 'cloud ethics' as an ethics from within—not from outside. She does not support the claims of transparency because she thinks that the ethical work should be carried out within the opacity of algorithms. She states, 'What I call a cloud ethics is concerned with the political formation of relations to oneself and to others that is taking place, increasingly, in and through algorithms' (2020, p. 7). Later, she adds, 'Just as Spinoza's ethics are ever in formation, forged through ethos, encounters, arrangements, and combinations, so a cloud ethics is located in the encounters, arrangements, and combinations through which the algorithm is generated' (2020, p. 165).

Amoore largely discusses machine learning in her book. In its last chapter (6), she analyses how 'cloud ethics could be brought into being in a world that is increasingly shaped through the arrangements of the machine learning algorithm' (2020, p. 157). She states (2020, p. 162),

> A cloud ethics alters the relation to a future that says 'here is a good enough solution' by revaluing the aperture as a moment of decision in the dark, a calculation of the incalculable. The strategy involves reopening the multiplicity of the algorithm, digging under the stories, and attending to the branching pathways that continue to run beneath the surface.

Amoore's book has made a huge impact. She even touches on incomplete representations, remarking that 'Amid the widespread moral panic surrounding the black box of the algorithm and the political demands for transparency, accountability, and explainability, in this book I make a counter case for opacity and the giving of partial accounts' (2020, p. 164). Responding to the presentations about the book in a forum, she underscored this point: 'It is my case in the book that although algorithms cannot be accountable in the sense of a clear-sighted account (e.g., an "opening" of the black box, or "explainable" algorithms), they can be called to give partial accounts of their conditions of emergence (p. 9)' (2021, p. 46). The consequence of this process is that the multiple individuals involved in the different steps should pay attention to the ethical dimension of every part of the process. This supposes, as noted earlier, the incarnation of an ethos or habit.

Moral habits are virtues, and the excellence or perfection of any being lies in virtue itself, as it embodies habits that facilitate good actions. Virtue was a particularly relevant concept for ancient Greeks, who believed virtue applied to everything, but first and foremost, it referred to human beings. Used to express quality, ability, excellence, honour, nobility, the term itself lay at the heart of the Greek ideal of a human being. Greek *Paideia* (education) was the creation of virtuous habits, while courage, moderation, justice, and prudence (*phronesis*) were highly regarded virtues.

In *The Laws* (I, 631c), Plato praised wisdom, temperance, justice, and courage, whereas Aristotle's *Nicomachean Ethics* is a treatise on virtue, where practical wisdom, justice, and friendship stand at the forefront. For Aristotle, the virtue of a good eye is to provide us with great sight, and a horse's virtue is to be good in itself and to win horse races, to be useful to riders, and so on. When applied to human beings, virtue is the state of character that makes them be good and do their work well. Humans have a calling or a task to do—the vital development of their very being—and therein lies their excellence or virtue. Happiness, a self-fulfilling activity, hinges on a life of virtues, which provide the means to attain happiness.

Because of their virtues, human beings behave in the right way easily, swiftly, pleasantly, and naturally. Life moves fast, and we do not have the time to choose or to analyse the best knowledge process at every instance. It would be psychologically unbearable to spend the whole day deliberating what to do and how to do it. We need a number of habits that facilitate rather automatic reactions, and, if we want these reactions to be right, we need virtues to support them. In a nutshell, virtues bring excellence to everything, enabling the proper behaviour for everything, especially human beings.

Aristotle drew a distinction between two kinds of virtues: intellectual and moral. Intellectual virtues help our intelligence achieve its goals, endowing it with the ability to perform its duties properly, that is, to seek the truth in the case of the theoretical intellect, and to point our free actions towards the practical truth, in the case of practical intellect. In turn, moral virtues push our will to what it ought to want, facilitating the good acts of the will and passions, so far as they are ruled by our will and our intelligence.

Aristotle viewed moral virtue as a state of character associated with choice, relying on a means, i.e., the means relative to us, the beings determined by rational principle. The word 'choice' implies that the primary act of moral virtue is the right decision—the choice to do what is needed for a good behaviour here and now—which presupposes the proper discernment on means and actions by practical wisdom. 'Choice' also indicates that the good action is wanted and elected in and by itself: intention matters. Virtues facilitate good deeds without doing away with freedom; on the contrary, virtues require a constantly free behaviour.

The most important moral virtues are called 'cardinal'—from the Latin *cardo*, 'hinge'—because all good habits hinge on them. Cardinal virtues include practical wisdom or prudence, justice, temperance, and fortitude. Prudence guides all other virtues, while justice rules social order. Fortitude and temperance harmonize people's inner order, as they are swayed by the dictates of their will and their passions. Thanks to prudence, human beings can understand and judge what they should do, whereas the role of prudence is to apply the right reason to action, which requires an appropriate will.

How do we acquire moral virtues? Aristotle believed that nature endows us with the ability to receive virtues, and that habits bring this ability into fruition and fulfilment. Habits become embedded in human beings when they repeat the adequate actions time and again, which means that people acquire virtues through practice. Virtues are not isolated: prudence connects them with each other to build a system—much like a linchpin that holds every virtue in place and together. For Aristotle, the work of prudence is personal as well as essentially free, and it may vary depending on the circumstances: what is prudent for one person may not be so for another. The coordination of free prudent actions leads to social coordination, as human beings cannot flourish in isolation—our fulfilment calls for a life with others. Virtue ethics is currently one of the most widely recognized ethical theories, as, while deontological ethics focuses on duties, and consequentialism, on consequences, virtue ethics zeroes in on virtues and personal character. Some authors argue that virtue ethics is unilateral because of its emphasis on the subjective aspect of the ethical relation (an 'agent-centred' ethics). However, the Aristotelian virtue ethics also considers the goodness of the act itself and its consequences.

Shannon Vallor (2016) calls for the creation of 'technomoral virtues,' for she regards the current impact of technology as something completely new that causes a deep transformation in the way people live and how the entire world interacts with technology. Technological changes unfold too fast for institutions to grapple with them, and the whole technological process grows increasingly opaque rather than transparent. Kant-inspired codes or precepts remain outdated, while a utilitarian ethics cannot foresee the consequences of a continuously evolving process. Vallor's proposal involves developing personal virtues specifically trained on our current situation and challenges. She states (2016, p. 10):

> The technomoral virtues, cultivated through the practices and habits of moral self-cultivation that we can learn from the classical virtue traditions examined in this book, are humanity's best chance to cope and even thrive in the midst of the great uncertainties and vicissitudes of technosocial life that lie ahead. This hope will only be realized, however, if these virtues are more consciously cultivated in our families, schools, and communities, supported and actively encouraged by our local and global institutions, and exercised not only individually but together, in acts of collective human wisdom. This is a tall order, but not beyond our capabilities.

Vallor looks for traits shared by Aristotelian, Confucian, and Buddhist ethics in order to formulate a global technomoral virtue ethics. She finds the following common features (2016, p. 44):

1. A notion of the 'highest human good' or thriving as the purpose of ethics;
2. The idea of moral virtues as 'cultivated states of character';
3. A roadmap pointing to the acquisition of virtues;
4. An understanding of what people are typically like.

Concerning the ways to develop the technomoral self, Vallor suggests taking the following steps (2016, p. 64):

1. 'Moral Habituation,' which is a process.
2. 'Relational Understanding,' because people are relational beings.
3. 'Reflective Self-Examination,' a Socratic teaching that leads to taking responsibility for our own actions.
4. 'Intentional Self-Direction of Moral Development,' viewed as an inner desire.
5. 'Moral Attention,' which refers to the ability to stay attuned to the moral dimension of situations.
6. 'Prudential Judgment,' the habit of examining scenarios and making the right choices in specific situations.
7. 'Appropriate Extension of Moral Concern'—'to the right beings, at the right time, to the right degree, and in the right manner'. (2016, p. 117)

Vallor finally introduces, explains, and supports her list of twelve technomoral virtues, which she regards as necessary to face the increasingly rapid, global, and unpredictable technological changes of our time (2016, p. 120). This list includes:

1. 'Honesty'—an exemplary respect for the truth.

2. ‘Self-Control’—becoming masters of our own selves by making the right choices.
3. ‘Humility’—taking stock and accepting our boundaries.
4. ‘Justice’—standing up for and adopting the right behaviour.
5. ‘Courage’—to prioritize morality at all times.
6. ‘Empathy’—feeling with others.
7. ‘Care’—an attentive and emotional disposition to meet others’ needs.
8. ‘Civility’—the willingness to ‘make common cause’ with others.
9. ‘Flexibility’—an adroit ability to adjust to changes.
10. ‘Perspective’—to view moral phenomena as parts of a moral whole.
11. ‘Magnanimity’—a generous disposition towards lofty aims.
12. ‘Technomoral Wisdom’—this virtue binds all the other eleven technomoral virtues together.

In the third part of her book, Vallor elaborates on the need for these virtues, diving into a number of challenges posed by the algorithmic process, social media, surveillance technology, robotics, and human enhancement technology. This author argues that we need to improve our thinking and acting, avoiding biases and building a solid moral personality and society. This means that virtue proves indispensable, and, as Aristotle points out in *Nicomachean Ethics*, it is fostered by:

1. Formal and informal education, which promote a critical attitude and activism since youth—we must realize that imperfect human beings transfer their flaws to algorithms.
2. A suitable legal framework and ad-hoc institutions that uphold ‘algorithmic accountability.’

Éric Sadin (2015, pp. 233ff.) calls for a ‘reappropriation’ of oneself, prudently drawing away from or even rejecting the AI domain—both individually and collectively. Julian Nida-Rümelin and Nathalie Weidenfeld stress the relevance of the development of a strong personality, stating (2022, p. 93):

> In the digital lifeworld of the future, personal strength is required more than ever before in human history. The education system must adapt to this. The imparting of knowledge and skills must serve the highest goal, namely, the strengthening of the adolescents’ personality. The focus must not be on the passive absorption of prefabricated material but on the active mastery of complex judgments and decision-making structures.

Conclusion

We stand at the threshold of a new world, a new life, a new relationship with technology. This breakthrough has its pros and cons. We must navigate it in a way that leads to the human and social good. We need, as Bertolaso and Marcos posit, a humanistic technology that promotes virtues, especially care (2023, pp. 84ff.). Vallor points out, 'naïve technophilia and reactionary technophobia are equally blind and unthinking responses to technological challenges' (2016, p. 219). Given the characteristics of the new technologies that are emerging, the necessary tool to manage them is personal virtue. Technology 'consumers' must rely on virtue, while engineers and businesspeople have to be virtuous and experts in ethics. They cannot just confine themselves to their specific tasks: they should be aware of the entire process and look at it with a critical eye. As Vallor warns (2016, p. 244),

> This will not be easy, since the lion's share of the planet's technological and scientific resources are controlled by liberal societies operating in cultures focused on short-term private gains, alienated from any shared conception of the ultimate moral goods to be realized by collective action in the long-term interests of the human family. Even the notion that there are such goods—as opposed to the immediate, subjective, and often arbitrary preferences of private individuals—is no longer readily granted in many such cultures.

The most urgent mission entrusted upon us today is to teach correct and virtuous attitudes towards technology to our young. The new generations will definitively need to possess all moral virtues. Vallor emphatically expresses this need in the last pages of her book (2016, p. 245), 'This book recommends a classical solution: an energetic (perhaps even desperate) collective effort to reinvest our cultures in the habits of moral self-cultivation and education for technomoral wisdom.' Sadin (2015, p. 261) adds that the most instrumental endeavour of our times is to submit contemporary algorithmic life to a critical appraisal. Thus, we need to improve our thoughts and deeds, avoiding biases and building a solid moral ground for ourselves and society at large. This means that virtue is indispensable. Hagendorff (2020, pp. 111–113) also recommends supplementing ethics with technical explanations and enforcing deontological guidelines with virtue ethics. He stresses the need to adopt a positive view of ethics as a driver to promote a moral character that showcases technomoral virtues. He states (2020, p. 113):

> Cultivating a moral character, in terms of virtue ethics, means to educate virtues in families, schools, communities, as well as companies. At best, every individual, every member of a society should encourage this cultivation, by generating the motivation to adopt and habituate practices that influence technology development and use in a positive manner. Especially the subject of responsibility diffusion can only be circumvented when virtue ethics is adopted on a broad and collective level in communities of tech professionals.

In a later article (2022), Hagendorff circles back to the fact that principles alone do not suffice and specifies which moral virtues should complement ethical principles. He pinpoints four 'basic AI virtues'—namely, justice, honesty, responsibility and care—and two 'second-order AI virtues'—prudence and fortitude. Virtues build trust and make social interactions more predictable. Hagendorff describes and justifies the adoption of these virtues, noting that second-order virtues counteract psychological or situational forces with the potential to neutralize basic virtues.

Drawing from Wiener, Amoore, Vallor, and Hagendorff, in this chapter I have argued that we need a radical change in our behaviour and the behaviour of political and economic institutions in order to make the AI system work in collaboration with us. This shift primarily requires enforcing the principles proclaimed by recent AI declarations, legislations, and codes of ethics, enhancing them with a real ethical conduct that will be achieved with the cultivation of specific ethical virtues.

References

Amoore, L. (2020). *Cloud Ethics. Algorithms and the Attributes of Ourselves and Others*. Duke University Press.

Amoore, L. (2021). "Response by Louise Amoore". In "Book Review Forum: Cloud Ethics. Algorithms and the Attributes of Ourselves and Others", The AAG Review of Books 9/3, 36–49

Aristotle. (1954). *Nicomachean Ethics* (Sir D. Ross, Trans. and Intro.). Oxford University Press.

Beck, U. (1988). *Gegengifte: Die organisierte Unverantwortlichkeit*. Suhrkamp.

Bertolaso, M., & Marcos, A. (2023). *Umanesimo Tecnologico. Una riflessione filosofica sull'inteligenza artificiale*. Carocci.

Bynum, T. W. (2004). Ethical Challenges to Citizens of "The Automatic Age": Norbert Wiener on the Information Society. *Info, Comm & Ethics in Society, 2*, 65–74.
Chesterman, S. (2021). *We, the Robots. Regulating Artificial Intelligence and the Limits of the Law*. Cambridge University Press.
Heidegger, M. (2013). The Provenance of Art and the Destination of Thought (1967). *Journal of the British Society for Phenomenology, 44*(2), 119–128.
Hagendorff. (2020). *The Ethics of AI Ethics: An Evaluation of Gudelines*, Minds and Machines 30, 99–120.
Hagendorff. (2022). *AI ethics and its pitfalls: not living up to its own standards?*, AI and Ethics 3, 329–336.
Madary, M., & Metzinger, T. K. (2016, February 18). Real Virtuality. A Code of Ethical Conduct. Recommendations for Good Scientific Practice and the Consumers of VR-technology. In *Frontiers in Robotics and AI*, Sec. Virtual Environments, Volume 3. Retrieved June 2, 2024, from http://journal.frontiersin.org/article/10.3389/frotbt.2016.00003/full.
Nida-Rümelin, J., & Weidenfeld, N. (2022). *Digital Humanism. For a Humane Transformation of Democracy, Economy and Culture in the Digital Age*. Springer.
Plato. (1952). *The Laws*. In *The Dialogues of Plato* (B. Jowett, Trans.). Encyclopaedia Britannica, Inc.
Sadin, É. (2015). *La Vie Algorithmique. Critique de la Raison Numérique*. Éditions L'Échappée.
Vallor, S. (2016). *Technology and the Virtues. A Philosophical Guide to a Future Worth Wanting*. Oxford University Press.
Vallor, S. (2024). *The AI Mirror. How to Reclaim our Humanity in an Age of Machine Thinking*. Oxford University Press.
Wiener, N. (1950, 1954). *The Human Use of Human Beings: Cybernetics and Society*. Houghton Mifflin; 2nd edn Doubleday Anchor.
Wiener, N. (1966). *God & Golem, Inc.* The MIT Press.

CHAPTER 10

Conclusion

Abstract This closing chapter wraps up the metaphysical, epistemological, and ethical analyses about AI carried out in all previous chapters. It readdresses the advantages and perils of AI, stressing the need to devise formal remedies and controls, as well as to promote a virtuous culture around it to ensure its steadfast orientation towards the human and social good.

Keywords AI advantages and perils • AI and morality • The AI Mirror

In several parts of this book, I have quoted Shannon Vallor's (2024) mirror metaphor to refer to the AI system. AI has gathered and continues to collect endless data about our thinking and our actions, thus turning into a powerful reflection of ourselves. Not all human characteristics and deeds, however, are or will be completely expressed in digital data—for example, our moral condition, feelings, emotions, or background will not be fully conveyed. Then, while the AI system is improving day after day, thanks to the increasing traces that we leave behind for it to pick up, it will never encompass the entire human reality. Indeed, it only captures present and past data, not the possibility of a future shift in our patterns. Yet, it has sufficient information to mimic our sometimes-unethical thoughts and

R. F. Crespo, *A Classical Philosophical Approach to Artificial Intelligence*, https://doi.org/10.1007/978-3-032-18462-7_10

behaviours, which will prove harmful if there are no substantial changes in them. Erich Prem also uses the metaphor of the mirror, but the other way around, to show how the digital domain is shaping us—with similar consequences, too. He refers to the 'digital humanism′ initiative and stating. 'In digital humanism, the digital becomes the mirror that impresses a picture of who we are on ourselves and lets us realize what we should change—about us, about technology, and about the lives that we live' (2022, p. 90).

The AI system is out of control, looking for power and economic benefits (see Edward Lee, 2022), and the only way to fix this is to change our undesirable actions and intentions, incorporating virtues into our lives. We need declarations, regulations, codes of ethics, and accountability tools to control AI, but these mechanisms must be reinforced by our actual moral conduct. The AI system is in fact dangerous, as many thinkers assert, although its danger lies in the attitudes captured in the data that we feed it—we are victims of ourselves. When I use the pronouns 'we' or 'our,' I refer to everyone: state officials, AI companies' owners, engineers designing AI tools, people promoting their use, users. Clearly, some individuals or groups have more impact than others, because they have a more digital connection, or because they intentionally try to influence or distort the image on the mirror in their favour. People more closely involved in AI design and deployment have a special responsibility and need to cultivate the technomoral virtues outlined by Vallor, particularly wisdom. Sound regulations and good incentives have to focus on them because the law promotes the enhancement of virtues. However, almost everyone uses data and needs to acquire or boost those virtues. As Vallor herself (2024, p. 142) puts it, 'Remember: the danger to our humanity from AI is not really coming from AI itself. The call is coming from inside the house; AI can devaluate our humanity only because we already devaluated it ourselves.'

The concern for care and service, for the welfare of both people and society, should provide the pillars to build a safe and collaborative AI system. Moreover, if we set economic and political goals that serve human integrity and life, AI can be a great tool to accomplish them. A symbiotic combination of good legislation, education, and virtues holds the key for promoting the AI that all of us should desire. The future lies in our hands, in our ability to merge technology and humanity, as the subtitle of the book promises.

References

Lee, E. A. (2022). Are We Losing Control? In H. Werthner, E. Prem, E. A. Lee, & C. Ghezzi (Eds.), *Perspectives on Digital Humanism* (pp. 3–7). Springer.

Prem, E. (2022). Our Digital Mirror. In H. Werthner, E. Prem, E. A. Lee, & C. Ghezzi (Eds.), *Perspectives on Digital Humanism* (pp. 89–93). Springer.

Vallor, S. (2024). *The AI Mirror. How to Reclaim our Humanity in an Age of Machine Thinking*. Oxford University Press.

Index[1]

A

B

[1] Note: Page numbers followed by 'n' refer to notes.

R. Crespo, *A Classical Philosophical Approach to Artificial Intelligence*, https://doi.org/10.1007/978-3-032-18462-7

The manufacturer's authorised representative in the EU is Springer Nature Customer Service Centre GmbH, Europaplatz 3, 69115 Heidelberg, Germany. If you have any concerns regarding our products, please contact ProductSafety@springernature.com

Printed and bound by CPI Group (UK) Ltd, Croydon, CR0 4YY
21/09/2026
02219995-0001